THE
BOOK
OF
DIFFICULT
FRUIT

THE
BOOK
OF
DIFFICULT
FRUIT

Arguments for the
Tart, Tender, and Unruly

KATE LEBO

PICADOR

First published 2021 by Farrar, Straus and Giroux, New York

First published in the UK 2021 by Picador
an imprint of Pan Macmillan
The Smithson, 6 Briset Street, London EC1M 5NR
EU representative: Macmillan Publishers Ireland Limited,
Mallard Lodge, Lansdowne Village, Dublin 4
Associated companies throughout the world
www.panmacmillan.com

ISBN 978-1-5098-7925-0

Grateful acknowledgement is made for permission to reprint an excerpt
from 'Unpour', by Shailja Patel, used by permission of Creative Time.

1 3 5 7 9 8 6 4 2

A CIP catalogue record for this book is available from the British Library.

Illustrations by Na Kim

Printed and bound by CPI Group (UK) Ltd, Croydon, CR0 4YY

Visit **www.picador.com** to read more about all our books
and to buy them. You will also find features, author interviews and
news of any author events, and you can sign up for e-newsletters
so that you're always first to hear about our new releases.

for Sam

If sweetness makes fruits desirable, there must also be sharpness: no rose without a thorn. The palate rejects blandness even when attracted by sweetness.

—ELISABETH LUARD,
in *The Oxford Companion to Sugar and Sweets*

Contents

////////////////////

A Note on Difficult Fruit

Fruit is how I best relate to plants—being myself an animal, and hungry. "You do not eat / that which rips your heart with joy," Thomas Lux writes. But of course you do. We the animals are pleased to fulfill fruit's highest purpose: after frost and flower and water and light, to be cherished, carried off, and gobbled—core flung from a window, pit hocked over a rail, seed shit in the dirt, under the sun.

In this book, fruit is not the smooth-skinned, bright-hued, waxed and edible ovary of the grocery store. Nor is it a symbol of goodness and kindness and productivity and the virtues of civilization. Here, fruit is difficult. Blackberries are invasive. Huckleberries can't be farmed. Medlars must rot in order to sweeten. Wheatberry dust is more explosive than gunpowder.

Consider the apricot. To a small child, apricot kernels can be toxic if eaten. To the sick, the same kernels might be medicine. To the cook, they are a secret source of flavor. To the consumer who knows no better, the pit and kernel are garbage to throw out—which suits the fruit best of all.

"Fruit of the womb," we say. Or "fruits of my labor." What precedes the bounty contained by these clichés—the thing obscured but not erased by the feast, the mechanism that brings fruit to fruition—is pain.

Each fruit in this book is difficult in its own way. Some are impossible to domesticate or tough to prepare. Some can be medicine or poison, depending on the size of the dose. Some are an acquired taste; some can't even be acquired. Here, their culinary, medicinal, historical, cosmetic, and cultural roles show us where what nurtures and what harms are entangled. Imagine a blackberry briar, forbidding and sweet, and follow me in—

A Note on Recipes

Recipes are rituals that promise transformation. They blend the precision of an instruction manual with the faith of a spell and, no matter when they were written, occur in the present. In this book, I use them to suggest additional ways to experience each fruit's qualities and difficulties. You can skim them, if you prefer. Doing so won't interfere with your reading experience.

Most of these recipes will transform a fruit by using a small number of ingredients and simple processes. Often, the cook will need more patience than skill. A few of these fruits require precise, repetitive preparations that will drive some people crazy and send others into raptures. You'll find out which type you are soon enough (though if you want to test it, start with persipan on page 126). Keep in mind that many of these recipes will adapt to other fruits. A thimbleberry kvass can be a raspberry or blueberry or mulberry kvass instead.

Some will require special tools. If you don't own a mortar and pestle, and you're interested in cooking, buy one.

It's a tool used by pharmacists and cooks, the perfect accessory for this book. Get a food scale, a quick-read thermometer, and glass jars. Make sure to have a preserving pan—heavy-bottomed, ovenproof, nonreactive cookware like a wide stainless-steel saucepan, enameled Dutch oven, or copper jam pan. Buy a yard of good cheesecloth from a fabric store. Look up local orchardists and farmers. Ask neighbors and friends for extra/strange/mysterious fruit, if you live in the sort of place where that's possible. Consider growing your own.

I measure ingredients in U.S. customary units and the metric system, switching back and forth according to convenience. For example, I find it much easier and more precise to calculate the fruit-to-sugar ratios of jam using metric measurements, but when I'm making less variable quantities of pickles and beauty products, cups and ounces are easier. Specialized cooking techniques are spelled out within each recipe.

I am a home cook. My methods and style are inherited and gleaned. I would not call my training formal. Nor have I felt the need for formal training. Everything I know about cooking comes from apprenticeship to master domestic artists, a cookbook habit, online food forums and blogs, my family, and years of fooling around in the kitchen. No matter how deeply I researched difficult fruits, the most essential information came when I gathered, cooked, and ate them. In essence, these recipes are invitations. Try this at home.

THE
BOOK
OF
DIFFICULT
FRUIT

A: Aronia

Aronia melanocarpa
Rosaceae (rose) family
Also known as chokeberry, black chokeberry,
barrenberry, chokepear

Aronia berries taste vegetal like a grass stem, then sour like a crabapple, with a tannic pucker that rivals raw quince and deep-purple juice that stains teeth like wine. Aronia's folk name—in many places still the name they're best known by—is "black chokeberry," earned by what you do when you eat them.

Member of the rose family and native to the Eastern half of the United States and Canada, aronia grows wild in good or poor soil, in meadows or bogs or roadsides or the edges of oceans and lakes, a cold-climate bush that loves full sun. It has been farmed in Eastern Europe and Russia for wine making, food coloring, and dessert flavoring since the mid-twentieth century, but only recently have American consumers shown any interest. Today, the co-ops and health-food stores where aronia might be found usually sell it as superfruit: nature's medicine, chewable by the handful if you can stand the taste.

Aronia berries are high in polyphenols, the phytochemicals with antioxidant properties advertised to prevent cancer and banish wrinkles, allowing us to imagine an invisible serum that can slake, slough, or soothe the rust from our oxidizing, time-sensitive bodies. Anthocyanins give the berries their deep color and antioxidant load, while tannins contribute a dry, overpowering taste so strong that most people cut aronia berries with milder fruit to choke them down. Tannins often accompany anthocyanins, which means darker fruit tends to be more astringent. To get the most nutritional benefit, take aronia straight. The operative rule here, as it was in my childhood home: the worse it tastes, the healthier it must be.

My mother's best scheme to convert me to her health rituals is a fruit smoothie. She starts with leafy greens in a blender, pours hot ginger tea over the greens to cook them slightly, then adds blueberries, two tablespoons of ground flax, and a banana, and blends until liquefied. The blender we use to make this drink shares the name of the hope we place in drinking it: Magic Bullet, or "something that cures or remedies without harmful side effects." I often confuse magic bullets with silver ones, which do something else— kill mercurial monsters efficiently.

Lately, I've replaced the blueberries in my mother's recipe

with aronia berries. I would believe they are three to four times healthier than blues even if their packaging didn't say so, because they immediately assert their potency: they stain everything they touch with a deep purple that's not like blood but reminds me of it. When I smash an aronia berry between my thumb and the kitchen counter, I make ink that doubles as a nutritional supplement.

My husband, Sam, tastes my aronia smoothie and grimaces. He wants to know what happened to my smoothie skills. You get used to the taste, I say. You start to like it.

I don't believe aronia berries will keep my cells unoxidized and disease-free, but I do believe that by preparing and ingesting these so-called superfruits and willing myself to complete this ritual every day, I have created conditions for good health. When I catch a cold, or when my immune system begins to attack me, it won't be because I have abused my body, but because I have one.

Every morning, I drink an aronia smoothie and gather my pills. One Tic Tac–sized generic for allergies. Two cellulose sacs, clear so I can see the beige vitamin powder inside. Tablets of green-tea extract to clear my head, and tryptophan to calm my inner weather. One gigantic blue-black pill (magnesium for smooth digestion) and one average-sized green (nettles to curb histamine). Two mustard-hued turmeric prophylactics that might reduce inflammation and might do nothing at all, but have no side effects, so why not? Finally, four brick-red pills, mesalamine, to forestall a range of symptoms from ulcerative colitis, an autoimmune disease whose treatment, and only cure, includes removing the colon.

When my depression or allergies or inflammation returns,

I will, through this ritual, relieve myself of responsibility. I did everything I could, I'll say. All I can do.

Lately, my medicines are working. My *compliance*, as doctors might describe my willingness to follow this smoothie-and-pill routine, has been 100 percent. Lately, I feel just fine.

When I feel well—on days when my body's only flaws are the ones dreamed up by my vanity—these pills seem worth the chore of swallowing them. The gagging. And the money. When I don't feel well, I can blame my lack of compliance.

Current evidence-based medical advice says that antioxidants like those found in aronia berries might improve health when ingested as a naturally occurring element of a balanced diet. Even then, the fruits and vegetables themselves may deserve the credit, or the benefits associated with antioxidants might be accessible only in concert with other natural elements in fruits and veggies. Supplements that isolate and concentrate particular antioxidants—as if they are pure ore extracted from the waste rock of fruit—have no proven benefits and may even harm. What we know for certain is what we've known for ages: a diet high in fruits and vegetables is healthier than a diet that isn't. The faith my mother gave me in food as medicine is well placed, but when it comes to superfruits we should all remain agnostic.

An alternative guide to food as medicine—one that lingers in our collective imagination despite a slew of evidence that contradicts its advice—is the Doctrine of Signatures. This ancient method of identifying the medicinal properties of the vegetable kingdom observes a plant's shape, color, odor, or habitat and relates these characteristics to the human body parts and conditions we imagine these plants resemble. Lungwort's mottled leaves look like sick lungs, which signals that it aids respiratory diseases. Eyebright's white-lashed flowers indicate that the plant will help sight. Stinging nettles echo the sting of allergies, so we use them to reduce our histamine response. Eat carrots to see in the dark; eat bread crusts to make hair curl. In Christian cultures, God formed these "signatures" to help people identify how to heal themselves, but the Doctrine of Signatures appears with or without that ideology, and with or without that name, all over the world throughout human history, from ancient Rome to English herbalists of the Middle Ages, from Renaissance physicians to modern-day Israeli folk medicine, from Native American medicinal traditions to ancient and modern Chinese medicine. This relation between plant and human parts is underpinned by a shared understanding between healer and patient that the human body ends not at our skins but extends into nature; that what is in nature is also in us. It is not so much a list of prescriptions as a way of seeing and making sense of how a world that harms our bodies can also heal them.

To many people today—and plenty in past eras—the Doctrine of Signatures is laughable. Mere pseudoscience. And yet I haven't stopped believing my eyes. The resemblance between small round aronia berries and pills, between dark-purple aronia juice and lifeblood—they make sense. A visual rhyme that reinforces my faith in aronia's hidden attributes, a way to look for a cure when I don't know how to cure myself. I know these berries aren't magic. I know they aren't medicine. But I feel better.

Superfruits might be sold for their medicinal characteristics; their high-density nutrients may promote good health and longevity; and their righteous glow has certainly been documented by advertising—but superfruit is a marketing pitch, not a medical term. Truly medical food, as the FDA defines it, is food that has been processed and formulated to treat a particular ailment, like malnutrition, and is administered under medical supervision. Medical food is not your doctor's advice to eat a balanced diet or consume superfruits to manage particular symptoms.

We eat superfruits anyway. Some of them are even tasty! But why do health-focused Americans eat acai and goji berries from abroad instead of aronia from our own backyard?

Exoticism sells. Plus, "chokeberry" is bad branding. Not only is the name unappetizing, but chokeberry is a common plant unremarkable in appearance and seemingly useless as food. The Potawatomi tribe of the Great Lakes has long known chokeberry's medicinal properties—in *Native American Medicinal Plants,* Daniel E. Moerman documents their use of chokeberry infusions for colds—but that knowledge wasn't valued by immigrant colonizers as they spread across the Midwest, and it wasn't passed down in any mainstream way to their descendants. Only recently have commercial marketers replaced the "choke" in favor of the more alluring "aronia" (from the plant's scientific name, *Aronia melanocarpa*). And only after "aronia berry" replaced "chokeberry" did marketers find eager North American customers.

Though aronia berries are also called barrenberries and chokepears, they are not choke*cherries*, a *Prunus* species with dark fruits that look like chokeberries but have pits like cherries. It is possible to confuse the two. They share the astringency and antioxidant load that gives both species their choke, but whereas the entire chokeberry plant can be consumed without consequence, parts of the chokecherry plant are poisonous. Do not eat chokecherry leaves or seeds. They contain compounds that the body turns into cyanide.

What I remember most about my mother's illnesses, whose symptoms were largely invisible because they came from migraines—"just another tired mom," one doctor diagnosed—are her diets, which hovered at the edge of our suppertime routines. Sardines, broccoli, fiber, flax. We did not have to eat those things but watched her eat those things, taking them into her body like a cure.

In my favorite childhood memory, it's 3:00 p.m. on a Tuesday the year I've learned how to read. I am sitting on the floor with a stack of Sears catalogues in my lap, sealed in a quiet that's like the quiet I feel right now, writing this.

I am reading. That's all.

When, as an adult, I first told my mother about this, she cried.

She cried because she was not there.

She was not there because, by the time she was thirty, the migraines that had felled her occasionally since childhood began to strike her down every week. I remember her home from work at the clinic, her bedroom curtains pulled against the sun, an ice pack clouding her forehead and eyes, the nausea and pain larger than she could support and stay upright. I remember a tunnel of bedclothes with my mother inside, still and quiet and breathing, waiting to be released.

My father was also not there. He was at a construction site. He's never said he feels sorry for this. Why should he? He would have felt guilty for not working.

My mother is sorry, my mother is sad, because every hour she retreated from us in pain is an hour stolen from her. She is still totaling that theft.

I tell her I was happy alone; her headaches gave me time to read and dream. This does not diminish her loss.

Mom says, still, year after year: This is the year I will get my health under control. One of the main methods will be a strict diet of ever-changing eliminations. She just needs to be good. She just needs to be better.

She is good. She is better.

And then she's not.

Sometimes I think she's trying to cure herself with diets because diets are impossible to follow perfectly, so when she slips up she can rest the blame for her illness on herself. Not God. Not science. *Compliance*. The regimen that will restore her health. The argument she makes to a body that won't listen.

"Don't give her that," my mother scolds. This month she's vegetarian. My father has handed me the best bit of fat off the roast he just pulled from the grill.

I put the meat in my mouth and chew, black sear on juicy fat and tender shreds of muscle. The best piece.

He looks at Mom with a tight smile that says this is none of her business.

"That stuff is going to kill you," she says.

I reach for another piece. I taunt her with my eating.

I am not a child when this happens. I am thirty, the age she was when migraines knocked her out. I feel a flash of fury like a child, though, because here again is the moving standard of health I cannot meet.

I will not give up fatty roasted meats, but I do drink a smoothie every day. I make my smoothie nearly undrinkable with aronia. Small purple seeds stud the inside of the blender. I imagine those pips in my digestive tract, clinging to me.

It is unpalatable to consider such things—how food is not just pleasure and nutrition and relief, but made of pieces that smear and stick to your insides, matter that your body has to work to expel.

I'm describing this as if it is a problem. As if some perfect gruel exists that moves through you cleanly, politely, a Gwyneth Paltrow dream-diet of ginger and air.

There is such a diet. On a low-residue diet, one must avoid any fruit with small seeds (especially berries), peels, fiber-rich substances, dairy, grain, and meat. It is a diet for when even homeopathic cures that are unproven to help but certain not to hurt—except my wallet—aren't helping. This diet provides the righteous difficulty of a cleanse while treating your insides like a toothless convalescent. Congee, boiled carrots, baked chicken. Gummy fuel for a gummed-up machine.

My GP is a naturopath, not an M.D. She recommends a lot of supplements. I also see a more traditional nurse practitioner at my city's gastroenterology clinic. "If you don't notice a difference within a week," my nurse practitioner says, "don't bother taking more. You'd be surprised how many of these natural treatments haven't passed clinical trials."

My naturopath agrees, but her timeline is different: she'd like me to put faith in subtle and constant intervention, try these homeopathic cures a little longer. "Think of your immune response as a bucket," she says. "The bucket can hold pretty much anything until it is full, and then it can't hold anything. Your sensitivity rises, even to substances that didn't bother you before." When I take nettles and turmeric or avoid dairy, it's because we're trying to empty my bucket.

"It's also a matter of what you consider an acceptable amount of irritation," she says. "One person's inflammatory response might be another person's no-big-deal."

"No amount of irritation is acceptable," my mother says when I tell her this. "Pain is not normal." I can hear what she's stopped herself from saying. Cut meat, dairy, sugar. Add greens and berries. Make the right choices now so you don't end up like me.

Sometimes, when I tell people about the physical therapy my mother does for a living, I tell them she's a good healer because she, too, is sick.

She manipulates her patients' fasciae and muscles, applies pressure in just such a way that they cease to hurt, then holds that position until she feels a pulse free itself and flood cramped tissue with oxygen, bringing it back into the circulatory system's usual route. This can seem as if, in the crowded room of your body, she's found your pain throbbing alone in a corner. She can do this because of science and study, but also because her pain-struck body knows, through long experience, how to see, where to touch, what to say.

After my mother scolds us for eating meat, we eat as much as we usually would, but in defiance. My father grows heavy with it. We get used to eating separate meals at the same table, or dining while Mom sips water and says it's fine, she already ate. We calm down, and the demands of her diet will feel less painful and less dogmatic as they become routine. I still worry that refusing to eat the same food, even for the sake of her health, frays her bond with us. That her cure denies connection.

During my latest autoimmune flare-up—a feeling of pressurized emptiness, like a bubble is blowing itself up in my stomach, then pain, like my guts are ripping open—I exile aronia smoothies and all other fruits that can't be peeled and seeded. Gluten, whole grains, dairy, they all have to go. I do this to calm my system but also to exhibit health-seeking behavior. So my doctors will believe in me. So my family will be patient with me. So I can say to myself I did everything I could. All I can do.

A month into this diet, my nurse practitioner tells me my hopes were misplaced: boiled meals won't hurt, but they won't cure me, either. My asceticism has been rewarded with stasis, not healing. This news is a relief and a disappointment, equally calibrated. Just "Oh." Just "Of course." Eat a balanced diet. Take my medicine. Call her if I have any problems.

Today, I am in remission again, back on aronia smoothies, and thinking of how the breakdown of our bodies can change the ways we are loved. I will never be completely healed, my doctor said. I should not hope for a cure, though I can hope—with medication and close attention to how food makes me feel—to feel healthy.

My mother disagrees. For thirty-five years, she's pursued the cure for pain. Why would she be satisfied with stasis?

"I will figure this out," she says. For me. For herself.

"I am closer than ever," she says.

I want to believe her. I want us to be well.

Aronia Smoothie

This smoothie is said to cure the common cold, warts, irritable bowel disease, fungal infections, jealousy, paranoia, dislike of children, cancer, acne, tuberculosis, traffic jams, comma splices, and existential dread. It does not cure can't-take-a-joke or do-I-look-fat-in-this, but may ease both conditions when part of a balanced diet. As you blend this smoothie, remember I am not qualified to make health claims.

Yield: about 20 ounces

2 packed cups raw, washed greens (beet, spinach, kale, etc.)

1 cup hot ginger tea (made with a ginger teabag or by adding a ½-inch slice of peeled gingerroot to hot water and steeping 1 minute)

1½ cups frozen aronia berries

1 banana, peeled and broken into large chunks

Juice of ½ lemon or lime

2 tablespoons ground flaxseed (optional)

1 or 2 cups chopped apple, pineapple, or melon, for sweetness (optional)

Toss the greens into a blender. Wilt the leaves by pouring the ginger tea over them (aids digestion, my mother says), and add the steeped gingerroot if desired. Then add the aronia berries, banana, citrus juice, and, if using, flaxseed and sweet fruit. Blend until smooth, adding more liquid or some ice as needed or preferred. Drink immediately.

Aronia Dye for Paper and Cloth

Every time aronia stains a countertop, I remember this use for the berries. The same dye prepared for cloth can be used for paper. Dye in that order—cloth, then paper—to save time reheating the dye. It makes a rich, warm purple or a warmish purple-gray, depending on the material. This recipe can expand to fit larger sheets of cloth and paper. Make sure to keep a 16:1 ratio of water to salt until you have enough to cover what you want to dye; increase the volume of aronia berries to match.

If you need to *remove* aronia juice (or any other dark berry juice) from fabric, try this: Boil a full kettle of water. Place a large metal bowl in a deep sink, and stretch the stained fabric taut over the bowl; a rubber band snapped around the lip works well to keep the fabric in place. Then, from a height that will feel too high (2 feet), pour hot water over the offending spot, arching your body away from the bowl to avoid scalding splashes. Repeat until gravity and heat erase the stain.

Yield: about 1 quart

Water
Salt
1 unbleached white cotton dishcloth (tired and
    stained cloths work well here)
1 cup aronia berries
1 sheet cotton- or mulberry-based paper (not
    wood-pulp-based paper, which might dissolve
    in the dye bath)

In a small stockpot or Dutch oven, bring 4 cups water and ¼ cup salt to a boil. This salt bath is called a mordant (derived from the French word for "bite"). Mordant helps dye "bite" cloth more deeply.

Immerse the cloth and reduce the heat. Simmer for 1 hour, then remove the pot from heat, rinse the cloth in cold water, and squeeze out any excess water. Set cloth aside and discard the salt water.

In the same pot, bring to a boil 4 cups water and 1 cup aronia berries (or more of both, to suit the size of your fabric, keeping a 4:1 ratio of water to berries), then add the cloth. Simmer for 20 to 30 minutes, then remove the pot from the heat and let cool for 1 hour, or up to overnight. Make sure the cloth is completely immersed in the dye. The longer it sits in the dye bath, the deeper the hue will be. The final fabric will be lighter than how it looks in the pot.

Once the cloth is at your preferred hue, remove the cloth and hang it on a clothesline to dry. Don't wring out excess dye—just let it drip onto the ground. Or wring it out if you enjoy the streaks of color this creates. The excess dye will stain whatever it drips onto, so if you're drying cloth or paper over a surface you don't want to stain, cover that surface. Rinse the dyed cloth in cold water until the water runs clear and any small berry particles that were clinging to the fabric have washed away.

When it comes time to wash aronia-dyed fabric, hand-wash it in cold water, or wash it separately in a machine on the cold-water setting. This dye will fade over time, and it will stain other clothing if you mix the load.

To save the dye, strain out the berries (you can still use them for smoothies, if you like; they're less flavorful and probably less nutritious, but still edible) and refrigerate the dye until ready for paper use—though not longer than 3 weeks (or whenever it starts to smell strange).

TO DYE PAPER, pour the cool dye into a bowl and dip paper into it. Hang the paper to dry. You can roll the paper, then dip one edge in order to create a field of color along that edge, or you can make stripes by winding rubber bands around the roll. Wherever the bands are wound, the paper will stay white.

B: Blackberry

Rubus armeniacus (Himalaya blackberry) and
Rubus ursinus (trailing blackberry)
Rosaceae (rose) family
Also known as bramble, bumble-kite, bramble-kite,
bly, brummel, brameberry, scaldhead, brambleberry,
fingerberry, dewberry

My childhood hill, where a wall of blackberries divided houses from what was not yet houses, smelled of fruit all August, a low, sweet note shot with mildew—jammy and muggy and fertile. To breathe deep was to be pierced with that scent. I stained myself purple eating the berries I could reach, but I most wanted the fat ones at the top of the briar. How could a plant be this sweet yet so thorny? Owned by my neighbors, yet so wild? Fences I could climb; blackberry brambles I could not. I entered the unbuilt land at the bottom of our subdivision through an opening in the blackberries and wandered the pines until they ended in a stranger's backyard or an animal emerged from the under-brush: a dog to chase me away, horses corralled by chain link. One day, the trail home was blocked by a pregnant goat too busy stripping leaves from blackberry vines to let me pass. I grabbed her lead and walked with her through the cul-de-sac until a woman in a white truck pulled up and yelled, "Hey! That's my goat!"

My family lived where the city wasn't yet the city, and our blackberries were Himalaya blackberries. A cultivar whose seeds were imported to California at the end of the nine-teenth century by Luther Burbank, the world-famous bot-anist, Himalaya blackberries (colloquially known today as Himalayan blackberries) love mild climates and land that's been staked, claimed, and torn up. The disturbed ground between the new house and the old road behind it, the ignored corners of the suburb. In western Cascadia, they rule abandoned houses, alleys, the front yards of people who don't maintain their front yards, any wedge of space that isn't yet valuable enough to be redeveloped, roadsides and riparian zones, acres of national parks.

They are the descendants of briars that Burbank called "Himalaya Giant" because he'd sourced their seeds from India. Released to the public in 1885 through his annual catalogue, these Armenian blackberries (not Indian, as Bur-bank had assumed), when transplanted from their native soil to the North American West Coast, became explosively productive. The USDA Forest Service's guide to managing them describes their growth habit as invasive, warning that canes can grow "up to fifteen feet tall before arching over to form trailing vines up to forty feet long . . . ribbed and reddish with cat claw–like thorns." Once they root, they're almost impossible to kill. Whereas modern blackberry cultivars such as Triple Crown, Obsidian, and Prime-Ark Freedom were bred to grow root systems in a compact ball, wild Himalaya blackberries spread their root systems wide, rhizomatically, then shoot up. When the tips of their canes loop back to the ground, they root there, too. They creep over native vegetation, crushing the understory—including native trailing blackberries, which germinate the same way in the same sorts of places but look flimsy next to a Himalaya blackberry briar. Some sources say they erode soil; others say they preserve it. They attract pollinators, which is helpful, but they also attract spotted winged drosophila, a vinegar fly that lays eggs on unripe fruit and matures into larvae while blackberries ripen.

A popular method of extermination calls for three straight years of poisoning the taproot while also cutting off the plant's canes. Another recommends burning the canes, a spectacle that the USDA says is "visually pleasing" but "not very effective," because fire isn't hot enough to kill the rhizomes beneath the soil. Herbicides can be trouble, because what's able to kill a Himalaya blackberry will probably also kill whatever it is choking. Goats can be effective. But the simplest solution is to keep blackberries trimmed and learn to live with them.

Burbank's early description of Himalaya blackberry fruit—not a berry, but an aggregate of drupelets embroidered to a core—calls it "small-seeded." What he does not say is how very many small seeds there are, and he makes no mention of the berry's hairiness—which is not hair, but hairlike remnants of the ovary that produced each drupelet. Their reputation for fresh-eating or jam is mixed. Some people insist Himalaya blackberries are not as sweet or deeply flavored as native or commercial blackberries, and some people can't stand their seeds. Others love their lush black drupelets, guard secret berry patches, and treasure the harvest as a summer ritual. Whatever the opinion, there's no denying that Himalaya blackberries are more abundant than other varieties, easy to find, and always free.

One day, while stuck in traffic, I watched a woman pick blackberries that were spilling over a retaining wall as she waited for a bus. She touched them gently before picking them, checking their ripeness. Himalaya blackberries darken before they ripen, as all blackberry species do, and will look ripe before they actually are. The best way to detect ripeness is to feel if they're ready to fall from the cane. Picking takes practice and patience. If the berries don't detach easily from their receptacles, leave them for tomorrow's forager.

Despite the thorns, blackberry pickers at commercial farms usually don't wear gloves because gloves dull our ability to detect ripeness. Fabric gloves also store and spread bacteria more than bare skin does, which is a problem for farmers who sell fresh berries to grocery stores. Latex gloves protect hands from berry stains, but they increase waste and must be scrupulously washed and replaced or they, too, will spread bacteria.

As for roadside berries, some people think they shouldn't be eaten, covered in exhaust and dust as they might be. But who am I to say? A spot of sweetness within the day's traffic seems like a good thing.

////////////////////////

When Burbank first sowed *Rubus armeniacus* in his experimental gardens in Santa Rosa, California, his intention was to use the hardy, prolific plant as parent stock
for thornless blackberry experiments, one of his many
attempts to create bigger, prettier, longer-lasting, disease-
resistant fruits, vegetables, and flowers. The stoneless plum.
The spineless cactus. The frost-proof orange. Plants should
be bred to the benefit of humankind, Burbank thought,
and what was beautiful and useful to us was better than
what was not. As he wrote in his 1907 book *The Training of
the Human Plant*, "in the crossing of species and in selection, wisely directed"—plant domestication sped up for the
new era—he had found "a great and powerful instrument"
that led to a plant's highest possible form.

One of Burbank's greatest contributions to plant breeding
(and thus our modern understanding of genetics and inheritance) was his belief that heredity was a "stored environment,"
meaning plants were as influenced by their environment as
they were by their genes. Introducing a cultivar to a new environment could encourage the cultivar to develop a better version of itself; hybridizing wildly different plants and growing
them in new soil allowed Burbank to select descendants that
exhibited the best traits of their parents. If the new species
bested native species, well, that was to humankind's benefit
overall. He was an early believer of Darwin's theories of evolution, and he used those ideas to create breeding methods far
ahead of his time. He didn't use the word "genes" to describe
the building blocks behind his crossbreeds. Instead, he talked
about the plant's spirit. He talked to the plant, too.

The Burbank thornless blackberry contributed to the development of modern thornless cultivars. Burbank's theories of nature versus nurture were ahead of their time. But, like the Himalaya blackberry's culinary reputation, Burbank's legacy is mixed. *The Training of the Human Plant* promotes his faith in eugenics, the movement Nazis used to justify genocide less than a decade after Burbank's death. The book collects his opinions about the best way to rear children (though he had none himself), using plant breeding as a metaphor for how to breed, select, and grow superior people. He thought that, "out of the vast mingling of races brought here by immigration," the United States could spawn an ideal populace via "constant cultivation and selection," which would do away with individuals he considered physically, mentally, and morally defective, "so that in the grander race of the future these defectives will have become permanently eliminated from the race heredity." Inspired by his methods for selecting superior cross-breeds of plants, he writes: "What, then, shall we say of two people of absolutely defined physical impairment who are allowed to marry and rear children? It is a crime against the state and every individual in the state." As if societies should sort citizens the way Burbank sorted daisy cultivars for their beauty or potato cultivars for their use, discarding those among us who don't meet "virile" or "moral" standards. In *The Training of the Human Plant*, a noxious work of supposedly good intentions, we see how it can be dangerous to compare people to plants. How, taken to their extreme logical conclusions, such metaphors become—as they always half were—inhuman.

By the time Burbank died of a heart attack in 1926, the Himalaya blackberries he meant to parent an ideal breed had escaped captivity and spread from central California to British Columbia. By the fall of the Third Reich, Himalaya blackberries had naturalized, a horticultural term describing a foreign plant that can reproduce on its own without help from humans. Once this particular plant naturalized, it out-competed plants who lived in ecological balance with others, colonizing native environments and replacing them with itself. In their first half-century on their new land, Himalaya blackberries were beloved because they were bigger, faster, and more prolific than native trailing blackberries. If they are beloved today, it is not because they are the best, but because they are everywhere.

I can't imagine my landscape without them.

Blackberry Mulch

I visit Unger Berry Farms in February because that's when Will Unger has time to talk. The Willamette Valley is sodden and muted under a bright-iron sky. The dormant rows of canes feel reflective and hushed, like we're walking at the bottom of a lake. I'm asking Will about Himalaya blackberries, which he does not farm. "People want to know why we don't just let them be," he says, "as if we could strike some kind of harmonious balance with them. They are weeds. Weeds are competitive. That's just how it is."

Will's family helped popularize the Hood strawberry, a highlight of early-summer Pacific Northwest cuisine and a loss leader at New Seasons Markets throughout Portland, Oregon. Priced as delicacies (but still cheaper than what they're worth), these strawberries attract customers and increase sales elsewhere in the store. They're known by their intense sweet flavor, delicate fruit, a red color that's consistent to the core, and by the green boxes they (and all Unger berries) arrive in, the ones that announce "Oregon Summer" in a blaze of yellow print.

Unger Farms also grows blueberries, raspberries, and blackberries on 143 acres in Cornelius, Oregon, where for the last 35 years they have combed their hills into rows of fruit bordered by road, stream, and forest. Along most of those borders, especially within the forest, Himalaya blackberries reign. Each year, Will and his family knock the brambles back but lose 1 or 2 feet of land to them anyway. One winter, Will attached a front loader to a tractor, flattened a bramble taller than he was, then mowed over it again and again until the pieces were too small to reroot. By the end of the battle, he'd gained 10 more feet of land. "The roots are still there, though," he says. "The blackberries will be back. It's like a horror movie for farmers."

Also in winter, the Ungers prune their 7 acres of cultivated blackberries, many of which are trellised so their fruit spurs—second-year growths called floricanes—hang in the aisle within easy reach, allowing pickers to harvest an annual 40,000 to 50,000 pounds of berries more efficiently while sparing them most of the thorns. How do you know which canes to prune? I ask, thinking of a Himalaya blackberry's impossible snarl. "There's usually a visible difference," Will says. "The stuff that fruited tends to have hardened off, like bark. It's usually a darker color, more of a purple. The new primocane"—the fresh, aggressive growth that will fruit next year—"tends to be pretty lush. It's actively growing in the summer and fall, while the old floricane just sits there after we harvest the fruit." Those of us who live with Himalaya blackberries can look for similar characteristics when training our brambles to bear fruit but not take over the backyard.

To prune his blackberries, Will and his team select the old canes and cut them off at the base. Then, while wearing welding gloves, they carefully unwind the canes from their trellises and throw the cuttings into the aisle behind them. After that, they drive a tractor-mower over the aisles, repeatedly mowing the canes until they're scattered in small pieces over the field. They leave the cuttings to mulch the plants they were just cut from. Then they begin the next task: trellising this year's primocanes before they become next year's floricanes.

When I ask Will how a person might use Himalaya blackberry brambles as mulch at home, he sighs, like he's already tired of thinking about it. "It'll be a challenge without a tractor," he says. "But here's what you could do."

After pruning blackberry floricanes, cut them by hand into foot-long lengths and arrange them into a low pile. Run a lawn mower over that pile. Rake the mowed canes back into a pile, fluff them up, then mow over them again. Repeat until the canes are less than two inches long. Spread them over the area you want to mulch. Or, to help them break down faster, rototill them into the soil. They won't reroot. "We mulch all the cuttings on the spot, because moving them is labor-intensive," Will says. The composted canes benefit the blackberry plants, but his main intention is to avoid the trouble of figuring out what else to do with them.

Blackberry Shrub

Sweetness in a berry isn't just sugar. It's sugar and acid. As the berry ripens, acid decreases and sugar increases. Too much sugar—as often occurs in overripe fruit—makes a dull-tasting berry. This blackberry shrub exaggerates and preserves the sugar-acid balance of a great berry. Also known as drinking vinegar, shrubs are an ancestor of soda pop, a sweet-sour beverage that will conveniently filter out annoying Himalaya blackberry seeds.

This method is a collage of the advice I found in Elizabeth Post Mirel's *Plum Crazy*, Sandor Ellix Katz's *Wild Fermentation*, the *White House Cook Book* 1905 edition, and the internet. Each source offered a different process but the same ratio—a 1:1:1 mixture of fruit, vinegar, and sweetener. Heating fruit with vinegar is the traditional way to make a shrub, resulting in a cooked taste— which isn't a bad thing, just not what I imagined for this shrub. A cold process preserves the fresh-fruit taste but takes longer. To preserve the high-summer flavors of Himalaya blackberries, I used a cold process.

To serve, fill a glass with ice, spoon 1 or 2 tablespoons of shrub over the ice, and top with sparkling water and lime.

Yield: 2¼ cups

    1½ cups (6 ounces) fresh blackberries
    1½ cups good apple-cider vinegar
    1½ cups granulated sugar or honey
    4 whole cloves, or 2 cardamom pods, or 1 cinnamon
        stick, or 1 thick slice fresh ginger (optional)

Combine all the ingredients in a quart jar and cover with a lid. Let sit in a sunny place, shaking the jar once a day to help the sweetener dissolve. Once it has (sugar takes about a week, honey is faster), strain the fruit from the shrub. Discard the fruit and spices, and bottle the shrub.

Or keep going. Pour the shrub over another 1½ cups blackberries, let it sit for a couple days, strain it, taste it, stop there, or pour it over another 1½ cups blackberries, etc. Stop when the shrub tastes as intense as you desire.

Shrubs can be stored at room temperature in a dark place indefinitely, but flavors will degrade when kept for over a year. Drink it before then. If you're unsure whether your shrub is still good, smell it. Does it smell like fruit and vinegar? Is its color what you'd expect it to be? If so, your shrub is most likely still good, its acidic environment having done its job of prohibiting bacterial growth and preserving fruit flavor.

C: Cherry

Prunus cerasus (sour cherry) and
Prunus avium (sweet cherry)
Rosaceae (rose) family

My mother's sister died of breast cancer when she was thirty-four and I was eight. I thought she'd caught the disease from eating maraschino cherries and hot dogs, a belief no one corrected because I never said it out loud. Eventually, in time that passed too quietly or too gradually to notice, I stopped believing she died from processed food. At no time did I stop eating maraschino cherries. This was, in a way, like smoking would be later: an abstract poison whose intake, because it had not yet killed me, made me immortal.

The story's often told this way:

The maraschino cherry as we know it was invented at Oregon State University during Prohibition to garnish virgin cocktails.

Real maraschino cherries were historically a Croatian delicacy made of Marasca cherries, a sour variety that's still grown and prized today. True maraschino liqueur is distilled from Marascas and their crushed pits, then combined with cane syrup and aged.

"Imitation" maraschinos have been made in the United States since the turn of the century, with a variety of nonalcoholic brines. In the 1920s, East Coast maraschino manufacturers ignored Oregon's and Washington's crops, because their Queen Anne variety went to mush when preserved. Enter OSU's cherry wiz: after years of experiments, Professor Ernest Wiegand discovered that adding calcium salts to cherry brine keeps cherries plump.

After 1940, the FDA decided it was lawful to call "imitation" nonalcoholic maraschino cherries just plain "maraschino cherries" in recognition that U.S. consumers by now assumed maraschinos were, as they are described by the FDA today, "cherries which have been dyed red, impregnated with sugar and flavored with oil of bitter almonds or a similar flavor." Modern maraschinos are bleached, preserved in a variation on Wiegand's sweet brine, and dyed with FD&C Red 40, a food coloring that has been accused of triggering ADHD in children. No study has conclusively proved that link.

In the United States, we usually derive the bitter-almond oil needed to make maraschinos from apricot kernels, not bitter almonds, unless we produce the flavor synthetically in a lab. True bitter almonds have higher concentrations of the chemical that creates almond flavor, amygdalin, than apricot pits do, but bitter almonds are harder to find and illegal to retail in the States because of their toxicity. A bottle labeled "pure" bitter-almond oil bought stateside is likely made from apricots. Some chefs I've known further clutter the nomenclature by calling apricot kernels "bitter almonds," which is what I now call the kernels I saved from last summer's harvest and stored next to my nutmegs.

The pits of cherries, plums, and peaches contain amygdalin, too, though in smaller concentrations, which means their almond flavor is not as strong. It remains common for cherry-pie recipes to recommend a teaspoon of "pure," "real," or "imitation" almond extract to gesture toward the bittersweet flavor of cherry pits without risking their potential poison.

Amygdalin is perhaps better known to some audiences as Laetrile or B17, notorious as a quack cure for cancer. At no time did my aunt consider amygdalin a possible cure, though evidence-based medicine didn't save her, either.

Reports vary on how many bitter almonds one could enjoy before regretting that choice. When amygdalin reacts with stomach acid, it splits into a nontoxic almond flavor called benzaldehyde and poisonous hydrogen cyanide, also known as prussic acid. A healthy human body can ingest small amounts of cyanide without ill effects, but higher amounts over time are poisonous. Toxicity depends on how large a person is, how healthy, and how concentrated the amygdalin content of the bitter almond. When I eat bitter-almond products—all made from stone-fruit kernels, not from true bitter almonds, and so slightly less dangerous—I do not eat them casually unless they've been baked. I'd prefer to eat persipan (a paste made from crushed apricot kernels and sugar that tastes bitter and satisfying, like dark chocolate) by the spoonful, but I eat it by the knifepoint instead. I do not serve the "almond" extract I make from crushed cherry pits raw to children, or serve it in quantities to adults as if it were amaretto, but I do add it in small amounts to pies. An advisory on these treats seems only right; some friends blanch at the flavors they're about to experience, fearful of cyanide no matter how small the dose.

To flavor jams, pies, and baked goods with stone-fruit pits, most recipe writers recommend blanching the kernels to be on the safe side. Others insist that blanching does nothing but degrade flavor, that the boiling point of water is 212°F and hydrocyanic acid does not dissipate below 354.2°F, so kernels must be roasted instead. This is how mahlab, a Mediterranean spice, is made. When crushed and sprinkled over pastries, it tastes like marzipan, but taken straight from the spice jar, it tastes first like pine nuts, then like bitter wood.

By kernels, I mean the nut-shaped lump you'll find if you smash the stone of a stone fruit. Beneath the kernel's wood-patterned skin is a white heart full of flavor. Cherry kernels are bead-sized and hard to extract whole from the crushed shell after the hammer's done its work, but this is just an aesthetic problem. To remove the skin, blanch the kernel in boiling water for ten seconds, let it cool for a moment, and squeeze it between your index finger and thumb until the skin slips free. Then, if you're feeling brave, eat it.

///////////////

In the house where I lived my last year with W, our neighbor's sour-cherry tree grew over the fence, which made it half ours no matter what she said. During a year of normal weather, the cherries would ripen the week of the Fourth of July. There was a five-day window when I could pick them and they'd be good. After that, they'd darken and soften and fall all over the back deck. Fallen cherries attracted birds, which attracted my cat, who was named after a cleaning tool—Swiffer—because he looked like the business end of a mop. He would attack and eat the birds and leave their feet and beaks by the door for us. This became, in our household, the spoils of midsummer.

One night, drunk out of his mind, one of W's friends chopped his own cherry tree down. It happened a week after his Independence Day barbecue, when we had been sitting under that tree, drinking beer and picking the sweet Rainiers hovering just above our heads. He did not compare himself to George Washington. Had he done so, I would not have felt any less like he'd just strangled his dog. "They made such a mess," he said.

W and I had cherry trees in our backyard, too. Overgrown Bing trees that bore sweet fruit, or fruit that would have been sweet if we had been able to reach it. The nameless cherries from the tree next door were far more neighborly, brushing our shoulders as we leaned against the back deck's railing, dense and sour and no good raw, no kind of quick treat for the picker. They made great pie.

At the time, I was learning my way around a pie plate. Baking seemed like a way to be powerful but nonthreatening; a confrontation with sugar was a fight I knew how to win. In addition to picking fruit and rolling crust, becoming a pie lady meant imitating a knowing voice when writing down practices I'd just learned from a cookbook or made up on the fly because I hadn't bought real equipment. The imperative mood was essential. In the absence of a genuine cherry pitter, the pointy end of a chopstick will do. Press the tip into the pit at the cherry's bottom, opposite the stem end. When the pit eases loose, drag it out by the stem. Discard the stems and save the pits. You can use them and the kernels within them to make almond-flavored simple syrup, vinegar, cream infusions, and jams.

I made those serving recommendations years before I made the foods I recommended. What mattered then was the gesture of knowing, practicing how it would sound. For some of us, this is how recipe writing starts. We try on the voice of the Cook Who Knows, offer likely answers, and build authority from there.

The best cherry-pie recipe is a simple one. When fruit is high-quality, our job is to preserve its natural virtues. Start with five cups of sour cherries, one cup of white sugar, a teaspoon of cherry-kernel extract, salt, and instant tapioca, mixed and poured into a double crust that's been glazed with egg white, dusted with coarse sugar, then baked in a hot oven. A thin layer of persipan spread over the bottom crust before adding the cherry filling adds a twinge of danger to this otherwise conventional dessert—or could. I haven't tried making cherry pie with persipan yet, but now that I've written that, I will.

In my new neighborhood, a family down the block has flawless sour-cherry trees on their fence line but for some reason does not pick the fruit. Sam and I decided to ask if we could. An old woman answered the door. Her husband had planted the trees years ago. He was dead now.

"I'll make you a pie," I said to her daughter, who translated my offer into Russian. In the foyer, a little boy was sprawled on his stomach, flushed and asleep on the bench where we would have taken off our shoes had we been invited in. "You don't need to make her a pie," her daughter said. "But you can pick the cherries." My cutoff shorts kept riding up as I moved. Her eyes followed my hand as I grabbed the cloth and pulled it down.

When we came back the next week, the cherries were perfect, too perfect, ready to fall and falling all over the sidewalk. We rang the doorbell to say hello. A new daughter appeared.

Her mother didn't remember us. The daughter wanted us to go away but would not say so. "I'm going to make you a pie," I said to the old woman. "No, not necessary," the new daughter said. "Really, I don't mind!" I said. "Maybe we can exchange recipes." The women did not scowl, but they did not smile, either. "Wash the cherries before you eat them," the daughter warned. "They've been sprayed."

While we picked, the boy who'd been sleeping last time ran out of the house with his sister and joined us beneath the trees. "Do you like cherries?" I asked the little girl. She refused to say. She grabbed a cherry from the dirt and threw it at her brother. A man in a suit pulled up in a minivan and said something in Russian that made them go inside and come back out with nicer clothes on. The rest of the family emerged from the house, dressed for church. "I'll bring the pie next week," I said as they passed by. The man smiled; the daughter scowled; the mother ignored us; they drove off.

The following week, our neighbor erected a chain-link fence around her trees. Branches still arched over the ugly metal, dropping all the cherries we had not picked onto the sidewalk. They rotted there; the birds ate them; we smeared them beneath our shoes and tracked them home. The mess was appalling. She didn't want a pie, I told myself. I didn't make her one.

One argument made on the third round of cocktails by an employee of an international fruit company to which I once sold my image, voice, and recipes is that this company, which might otherwise support legislation for GMO labeling, was instead against it for fear that clear labeling would also have to list the naturally occurring chemicals in non-GMO produce. The public would see the arsenic, the solanine, the cyanide, and become terrified of fruit.

Oxalic acid, for example, is found in buckwheat, spinach, sorrel, and rhubarb. Rhubarb recipes often warn to stay away from the leaves because they contain poisonous amounts of oxalic acid, but they never mention that rhubarb leaves taste terrible, which mitigates the danger of accidentally eating a dangerous number of them. According to *Bon Appétit*, a 130-pound woman would need to get past the taste of ten bitter pounds of rhubarb leaves to have an adverse reaction, which could include burning mouth and throat, nausea and vomiting, diarrhea, cardiovascular failure, coma, and death.

If we can't blame maraschino cherries for my aunt's cancer, what should we blame? TV waves? Anxiety? Rage? Three years after she died, when I was eleven, my mother took me to the pediatrician so I could show her my right nipple, which had swollen unnaturally larger than my left. I was terrified I had breast cancer, sure I was next. "It was puberty, not cancer," Mom remembers, "but I thought you would feel better if a doctor said it."

Years after my aunt died tragically young (but not from eating maraschino cherries), years after I left W and our fruit trees, I started a new July ritual: make maraschino cherries the old-fashioned way, free of food coloring and calcium salts, preserved for cocktails year-round.

I start with four or five pounds of fresh sour cherries and use a real cherry pitter now, a contraption that screws onto a standard canning jar and catches the pits. Sam and I drink Manhattans while we pit, served his favorite way: on the rocks with the previous year's maraschino cherries, never citrus. I combine the pitted cherries with a bottle of Luxardo maraschino liqueur and heat them briefly, because I read somewhere I'm supposed to do that. Then I seal them in jars with some of the pits to add a dash of benzaldehyde charm. Over the rest of July and part of August, the maraschinos' candy red will fade to a serious mauve. That's how we know they're ready, globes of liqueur-logged fruit that taste of what they once held safe at their centers. Summer glow and fair warning, true cherry and almost almond, the promise and poison from deep in their seeds.

"Almond" Extract

The first time I sampled a spoonful of this extract, my heart raced and my vision darkened. Of course, I was alone in the house. I called Poison Control, which had never heard of cherry-kernel poisoning but could say that if I was suffering cyanosis my fingernails would be blue and my face would look gray within 2 hours of ingesting the dose. I did not have these symptoms. I had 90 minutes to wait to be sure, but I did not need to wait. I was fine. I'd suffered a psychosomatic reaction to my fear of poisoning, and the cure was talking to Dale at Poison Control.

To understand how to weigh the dangers of cherry kernels, I called an expert at a research university who provided fantastic details but ultimately requested I not name them or their university. Benzaldehyde can be made in a lab without cyanide; the resulting imitation almond extract is a strongly flavored product that carries no risk of poisoning. As far as the university was concerned, my recipe unnecessarily courted danger, and they wanted nothing to do with it.

The European Food Safety Authority reports that a lethal oral dose of cyanide ranges from 0.5 to 3.5 milligrams per kilogram of body weight. The National Capital Poison Center notes, "The amount of amygdalin in the seeds of stone fruits varies widely, both between different types of stone fruits (e.g., cherries vs. plums) and even within the same type of fruit (e.g., cherries from one tree vs. cherries from another tree or geographic location)." My source estimated (reluctantly, because I kept asking them to) that a person could get a toxic dose from eating about two hundred raw cherry kernels in one day, which they said equals approximately 1 ounce of cherry kernels. The reaction will depend on weight, overall health, and immune-system strength. Problems usually begin in the bowels, where the body breaks down cyanide. When consuming raw stone-fruit extracts, use caution. A small spoonful will probably be fine (I've never suffered any ill effects), but don't treat yourself to a couple jiggers.

In this recipe, vodka draws benzaldehyde from crushed cherry pits to make an almond-scented-and-flavored extract usable in baked goods, but it also creates benzaldehyde's fraternal twin, cyanide. This extract is meant to flavor cherry pies and whipped creams, things like that. Do not use this extract as a cocktail ingredient, and do not drink the entire yield of this recipe—or even a shot of it—in one sitting. The maximum usage I feel safe recommending is one or two teaspoons for an entire pie or batch of cream.

In my research on stone-fruit kernels, I've seen many people advise blanching the kernels for a minute or two to boil off the cyanide. This is likely a myth. Hydrogen cyanide breaks down at around 354.2°F. Boiling water only reaches 212°F (at sea level), so blanching probably does not remove toxicity from the kernels—but, again, chemical reactions at different temperatures are highly variable, so I can't consider this the rule. I can be pretty sure that, if I use this extract in a pie, the cyanide will break down during baking. The benzaldehyde, which is what I'm really going for, is safe to consume and will remain to flavor my dessert.

I attempted this recipe with uncracked shells. It works, and it produces a nontoxic extract, but the almond flavor is weak. I also tried it with kernels only. This also works, but also not as well, plus it's annoyingly labor-intensive. The best method, which results in the strongest almond flavor, is to crack the shells with a mortar and pestle and include both shells and kernels in the extract.

Yield: about 10 ounces

1 cup cherry pits, scrubbed of any clinging cherry fruit
   and dried
About 1½ cups vodka

With a mortar and pestle, smash the pits in small batches to crack the kernels from the shells. No need to smash further once they're cracked open. Be prepared for flying shells.

Spoon the cracked kernels and shells into a clean glass pint jar and fill the jar with vodka. Cover it with a lid, put it in a dark place, and forget about it for a while. The "almond" extract will be almondy within 3 months, usable at 6, but best after a year or more. It is shelf-stable, like any extract, and will continue to improve with time. I use it by spooning extract off the top, leaving the pits to continue their business at the bottom of the jar.

Real Maraschino Cherries

The hardest part of this recipe is finding sour cherries, picking them, and pitting them. Everything else takes 30 minutes. I've used frozen (then defrosted) sour cherries for this recipe, but fresh cherries are better. Reserve the pits to help flavor the maraschino cherries and to make "almond" extract (preceding recipe). To preserve the shape of the cherry and your sanity as you pit pounds and pounds of them, get a cherry pitter.

Yield: about 72 ounces cocktail cherries

4½ pounds (about 2 kilograms) sour cherries, stemmed and pitted (about 11 cups)

750 milliliters Luxardo maraschino liqueur (one bottle, or 3¼ cups)

20 to 40 cherry pits

Prep nine 8-ounce jars by washing them and then drying them in a 250°F oven. Remove them when dry—they do not need to be hot for canning. Wash the lids and bands, and set them aside to dry. Keep the oven on.

In a stockpot, over medium-high heat, combine the cherries and liqueur. Bring to a soft boil, then immediately remove from the heat. Spoon the cherries into the jars until there's about ½ inch of headspace left (don't overpack—there may be some leftover cherries), add two or three or five pits per jar, then spoon the liqueur over all to cover, once again leaving about ½ inch of headspace. While the cherries are still hot, place the lids and screw bands fully onto the jars so they are fingertip-tight (tight, but not so tight you need strong hands to reopen them). Heat for 15 minutes in the oven, then remove and let cool. Check the seals as they cool. If any don't seal, use those jars first.

Their alcohol content will prevent spoilage, but processing makes double sure that the cherries you preserve in July will make a year's worth of Manhattans.

Store out of sunlight at room temperature. You can use them immediately, but they'll taste better the longer they steep. Store opened jars in the refrigerator.

D: Durian

Durio zibethinus
Malvaceae (mallow) family
Also known as the king of fruits

If it is a book of difficult fruit, it must include durian.

When I tell people I am writing about difficult fruit, durian is often the first they suggest. They have seen the "exotic food" lists, the "Top Ten Weirdest Fruit!!" clickbait, the headlines that sensationalize durian's strong scent. So, when I say "difficult," the fruit that comes to mind is this exceptionally odoriferous one.

//////////////////////

Instinct tells us not to eat the bad-smelling thing. It is unknown—or unliked. It could be rotten.

"Would you please, please, please, please, please throw that out," Sam pleads, somehow knowing that whatever the powerful smell is, I'm responsible for it.

Durian is grown throughout Southeast Asia, where Thailand is its biggest producer and Indonesia outlaws it on buses and trains. Durian trees can grow four times as tall as a telephone pole. They bear fruit that's nearly as heavy as a gallon of milk, with soft flesh armed in dramatic spikes that make handling difficult without gloves. When English speakers say "durian," they're using the Malay and Indonesian word for it, which is rooted in *duri*, Malay for "thorn." Falling fruit can kill. One must not loiter under the durian tree.

My first taste of durian was as candy, a beige lozenge with a slight pink blush that my boss at the time dared me to try. I unwrapped it from blue plastic and dissolved it at the roof of my mouth over the course of an interminable staff meeting. It tasted of strawberries and old garlic. I had to will myself to finish. My boss would use my failure to eat the bad candy as further proof of how I made her job difficult. I could not allow that.

My second taste of durian was at dim sum in New York City, visiting a man who would never love me. The durian was stewed, sweetened, and crenellated with flaky dough. We goaded each other to see who could enjoy it more, as if extra bites could prove who was worldly, who had better taste. The durian was like peaches laced with onions, and had a richness that made my chest tight. Each bite was a dare. Could I keep going? Prove I was game? He watched me eat the durian and I watched him eat the durian, and when the last bite arrived, soft and serene in its cracked coat of pastry, I let him have it.

All other attempts to enjoy durian—I have lost count of how many—were undertaken to write this chapter. If it is a book of difficult fruit, I thought, it must include durian.

Durian is easy enough to find fresh in the States if one lives in a metropolitan place with the sort of grocery stores that sell unusual imported fruit. The markets in Spokane sell six kinds of wild mushrooms and twenty varieties of apples, but to find durian I must visit one of the four Asian groceries in town. I guess this is the beauty of the global marketplace: when curious about a fruit to which I have little personal, historical, or cultural connection, I can get in my car, drive a mile down East Sprague, and buy some.

I'm out of luck finding fresh, rind-intact durian—which is not jackfruit, the fruit I've been mistaking for durian, as some durian neophytes do. Jackfruit's bumpy rind and intimidating size resemble durian's, but once you see the two side by side, durian's spikes are unmistakably spiky. The vacuum-sealed packs of frozen pale-yellow durian flesh are more expensive than I was expecting—$10 a pound to jackfruit's

$3—but I only need a cup or two. Down the aisle, the shiso I was looking for last summer, when I was making ume-boshi, has appeared. I've been fermenting ripe and unripe shiro plums—the closest I could find to ume—for the last five months in my basement. They need at least another seven months, so maybe it's not too late to add the shiso, which will stain them magenta and give them a sharp herbal scent.

The woman at the checkout is neutral about my shiso bunches and green-tea candies and bags of rice noodles, but says "durian" and catches my eye with a smile when she slides the frozen package over the scanner. I ask her if they ever carry fresh durian and she says no. "Just frozen," she says. "It's special. And very good."

When I buy durian, persimmons, or any other fruit that's not so obvious to a palate raised on apples and ba-nanas, she checks in with me about my purchase. "Have you had this before?" she asks. "Do you know how to eat it?" Confused costumers sometimes try to return the tan-nic or strange-smelling fruit, she tells me, complaining there's something wrong with it. "There's nothing wrong with it," she says. I hear the hint behind the helpfulness: if I have problems with this fruit, they're my problems.

Of course, she's right. What I don't enjoy about durian is my loss. Why should she allow me to make it hers?

I promise not to eat the persimmons before they're soft. I ask her how she'd eat the durian. "One of my customers says to thaw the durian, add twenty percent ice, and blend it. Then eat the durian like ice cream." She hasn't tried this yet but thinks it sounds good. "At home, I just eat it straight," she says.

///////////////////////

As the durian thaws, a faint scent appears, but only within a few inches around it; it's sweet and trashy, like a cantaloupe that's been left in the car. To get a good whiff, I have to cut open the plastic.

Fruit and sewers first. Then, if I leave the room and return, an industrial smell that's a little like turpentine. It reminds me of our garage, how we can't scrub the scent of someone else's motor oil out of the cinder blocks. The odor feels close and unpleasant, like a belch.

It tastes savory, then a strong peach or banana or something fruit flavor, then savory again to finish, with a texture that's creamy and softly stringy, not juicy, rich in a way that reaches down my throat and makes me feel like I've drunk a spoonful of heavy cream. For a long time, the aftertaste is of dessert that couldn't quite conquer the garlic I had for dinner.

I confess to you as I confessed then to Sam: each spoonful is harder than the last.

Those who have a taste for durian report lesser degrees of stink. Not, as Ernest Small's encyclopedia of *Top 100 Exotic Food Plants* records, "French custard passed through a sewer pipe," but a "sensuous banana pudding with a touch of butterscotch, vanilla, peach . . . and almond flavors and a surprising twist of garlic or onion."

Look, the "difficult" of difficult fruit is not code for "exotic" or "ethnic" or "other." To call a fruit difficult because it is unfamiliar to me and I do not like it would be ridiculous. A failure of imagination.

What do durian fans experience that I can't train or will myself to experience? What is durian like when you don't mind the onion burp, the melon rot, the turpentine perfume?

I would like to think I can teach myself to have a taste for anything.

Stink *can* be difficult, depending on whether one has a taste for something. One-thousand-year-old eggs. Gefilte fish. That rotting Icelandic shark-fin thing. Sauerkraut. Fish sauce. Garlic. Cheese!

It is a bit embarrassing, in a book of food, to admit the limits of my own taste. I have tried not to cloak my shortcomings in an armor of facts. And anyway, who would believe me if I suggested I'd never been disgusted by unfamiliar food? If I said I have transcended my childhood preferences and learned to like durian as I like canned cheese and turkey pinwheels? Or as I now like the blue cheese I mold in my own basement?

Sometimes when I run aground on these boundaries of taste, I think of a line from an Amy Hempel story: "At the smell of the dinner frying in sesame oil, the man's face changes. Mrs. Hatano reads the look as Other People's Food."

The man's expression is involuntary, which makes it revealing. Mrs. Hatano notes his discomfort and continues her business, cleaning the house and cooking a meal after the man's wife has died.

I do as instructed and blend the frozen fruit with ice, adding just enough water to keep the blades moving. I pour the pale-yellow slush into a parfait glass and spoon it into my mouth like ice cream. Some falls from the spoon to splat on the countertop and spatter my chest. Next time, I'll use a straw.

The durian "ice cream" melts over my tongue and coats my gullet with the familiar heavy sensation that accompanies all rich food for me, a pressure from mouth to chest. A gentle feeling, but not pleasant. A sensation that Sam, for whom nothing is too rich, says he cannot understand. As he watches, I eat until I feel sick. Just five bites! Then I turn to the fruit I have failed to love, whose smell is driving my husband insane, and I do myself a favor. I throw the durian out.

Durian Ice Cream

In the course of editing this chapter, I ate a lot of durian to make sure my particular adjectives for the way I dislike it (motor oil, onions) were not exaggerations. I ate durian fresh, frozen, freeze-dried. As pastry filling, as pudding, as ice cream. One day, while I was sharing a durian crêpe with a friend, she surprised me by jumping from the table, rushing to the bathroom, and spitting her bite into the toilet. What was happening? Could the crêpe possibly taste that bad?

Wait—when did I cross the durian line? Why didn't I notice? And did my friend have to be so theatrical about it?

"Kate, it's *disgusting*," she said, and began brushing her teeth.

I thought the durian crêpe was . . . good?

I used to feel my friend's distaste. Now I only observe it.

Then, finally, in my third year of searching for fresh durian in Spokane, I found one. Once again, the woman I bought it from had a lot to say about my purchase. "Expensive," she warned. It *was* expensive—$26 for a half-sized durian. That's okay, I said. "Are you sure?" she asked. I'm sure, I said. "Do you know how to cut this?" she said. Yes, I lied. Fresh durian is encased in a tough, spiky rind that cushions its long fall from tall trees but makes the fruit painful to hold and hard to hack through. To carry it to the checkout stand, I had to cradle it in my sweater.

"Are you sure you know how to cut this?" she asked again. I told her I wasn't sure. "Leave it outside until it softens," she said, "even though it's going to freeze tonight. Then cut it at the natural seams." As I handed her my money, she asked me one more time, "Have you had this before?"

It was cute, the durian. I hadn't anticipated that. Each time I saw it sitting by the back door, I wanted to pick it up and hold it like a cat. Over the next 2 days, the rind softened as promised. I cut along the seams with a cleaver, slipped my fingers around each side of the cut, and yanked the rind open. From there, the fruit was easy to scoop out. It was pale yellow, with soft lobes of tissue and large brown seeds that looked like chicken hearts.

Durian can, of course, be eaten straight from the rind. But it can also be sweetened with a little condensed milk, scooped into spheres, smothered with more sweetened condensed milk, and topped with a sprinkling of coconut. With these additions, the taste, for some, will be creamy, nutty, even a little chocolaty—and totally, arguably delicious.

Yield: about 2 cups

16 ounces frozen durian fruit
¼ cup sweetened condensed milk
¼ cup chopped frozen shredded coconut (optional)

Chop the frozen durian into 1-inch chunks. Put the durian chunks and sweetened condensed milk into a blender. Blend on low speed at first, so the blades don't fling the durian to the sides of the blender without blending them first. As the durian chunks break down, increase the blender's speed slowly to a medium setting. Let the blades whirl until the mixture is completely smooth.

Serve in chilled cups, and top with shredded coconut or more sweetened condensed milk. Or pour into an ice-cream maker and churn according to your maker's instructions. Freeze any leftovers. It keeps remarkably well, better than milk-based ice creams. You can defrost it and eat it chilled, and the texture will remain pleasing.

Durian Lip Balm

There's a long tradition of using fruit to scent, color, and flavor cosmetics, and of naming lipsticks after fruits that evoke a particular fruity shade. Those who love durian can wear that love right beneath their sniffers via this lip balm, which mixes real durian fruit into a simple balm recipe. Durian does not contribute color—expect this balm to be white in the pot, translucent on your lips—but it does smell like fresh durian. Some people will say this lip balm stinks. No kisses for them.

Yield: about 2 ounces

> 1½ ounces (about 2 tablespoons) fresh or
>   thawed durian pulp
> 2 tablespoons plus 1 teaspoon sweet almond oil
> 1 tablespoon unrefined shea butter
> 2 teaspoons beeswax pastilles

Mash the durian at the bottom of a small jar, and pour the sweet almond oil over it. Submerge as much of the durian as possible in the oil. Cover the jar, and let it sit in a sunny place for 2 or 3 days. Then strain the durian from the oil. You should end up with about 2 tablespoons oil. Discard the durian. (Or—reserve 1 tablespoon of durian and mix it with the ingredients in the next part of the recipe. This will make a more intensely scented product, but the fresh fruit may cause the balm to eventually spoil.)

Gently heat the strained oil, shea butter, and beeswax in a small saucepan, stirring to combine. Immediately after the shea butter and beeswax have melted, remove the pan from the heat. Pour the balm into a small glass or metal container, and let it cool.

Wash your cooking utensils in hot water right away, before the lip balm residue cools. Store at room temperature.

E: Elderberry

Sambucus nigra (black elder) and
Sambucus cerulea (blue elder)
Adoxaceae (elderberry) family
Also known as elder, common elder, pipe tree,
bore tree, European elder

I tend to think of plants as either landscaping or food and don't notice them until they're labeled for me, after which they appear everywhere.

Take elderberries. I was in my parents' new house, helping them prep my grandfather's funeral supper, when I noticed a spray of blue a hundred yards beyond the kitchen window. My father was still answering questions like "What was important to your dad?" with statements like "Being the center of attention." One of my uncles wasn't answering my father's phone calls at all. My mother was roasting another pan of vegetables. My brother was waiting for the mood to lighten before he shared his good Grandpa memories. My great-aunt was on her way from Fort Worth to give everyone unconditional love. And there, a short walk away, were the elderberries I'd been looking for all summer.

I never knew how small elderberries were, or how dark their juice, or how they cannot be made delicious, not really, not like a recipe for elderberry pie would have us believe. Even when they're doused with sugar, something chemical remains, some twinge where my tongue attaches to my mouth. A pinch in that soft part, like a warning.

Elders like the edges of things, how the soil drains better at the lip of a drop-off or the crest of a hill. They grow all over Europe and North America in wild places, roadsides, and yards, wherever there is cool weather, and are considered by some to have magickal properties. Their flowers are delicious gently battered and fried, or added fresh to baked goods. Essence of elderflower can be extracted by simple syrup or neutral grain spirits to make cordials and liqueurs, or extracted by water to make a sweet-smelling facial astringent. The flavor and scent of elderflowers are pleasures unequaled by almost any other flower, but remember, if you harvest all the elderflowers, you will have no elderberries.

Elderberries are used in jellies, jams, juices, and sauces. When crushed, they make an organic dye historically used by butchers to mark meat. Early-American settlers ate elderberries and used them as medicine, but, then and now, elders have seldom been cultivated. If you do not know whether elders are bushes or trees, this confusion is deserved. They are bushes that grow twenty to thirty feet tall and can be trained—by purposeful pruning or grazing deer—to look like trees.

Blue elders are distinguished from black elders by the silvery coating on their blue-black fruit. They're native to every place I have ever lived, but I never saw them until just that moment, when I was washing pans for the meal we'd come home to after burying my grandfather, the day after I wrote in his obituary, as requested, that he had three children, not five.

While he was alive, I didn't think much about my grandfather's being a pharmacist. I liked seeing him in his white coat, but it didn't occur to me until later to be interested in what he did behind those high counters. The only son of a railroad machinist, he knew early on he wasn't cut out for his father's kind of work. Pharmacy offered an upwardly mobile profession and a respected place in the community, though he never earned as much as he expected to. Eventually, my grandparents owned a small-town drugstore in Maquoketa, Iowa, in the decades before Walmart and Rite Aid made such a store an antique. My grandfather seemed to love work but never talked about work, and passed that habit down to my father. My grandmother managed the gift shop, accounting, and employees. They called their store Snow White Pharmacy. They'd inherited the name.

After my grandfather died, I took things from his boxes I didn't have permission to take. Baby photos of me and my brother. Snapshots of my missing aunts. My job was to construct a timeline of my grandfather's life without including photos of his daughters. I had given myself this job. My mother, to protect my father, had asked me to exclude the girls; my father, to protect his mother, agreed. When my father stayed too long in the company of these boxes, he grew irritable and tired, and my mother, previously supportive of my efforts, grew protective of him. I'd learned to wait to open the boxes until they were not around.

In a jumble of ephemera that looked like it was swept off a counter, I found a letter from my aunt Julie to my grandparents, sent in 2008. Julie had sent a letter like this to my elder uncle, too. "Never did get to the point," my uncle had said. "Spent the whole time repeating how happy she was." In my grandparents' letter, she writes around her subject, too, but gets direct in the last paragraph: "I will always consider you my parents," she wrote.

And what, in the end, did my grandfather consider Julie and her sister? His prodigal girls? His wild daughters?

After reading Julie's letter, I felt a jolt of what my father must have been feeling—I couldn't bear these boxes, either. But before I stopped looking, I took an old bottle of morphine and belladonna that was archived in the same counter-sweep as Julie's letter. "Probably inherited from the previous Snow White owner's pantry of drugs," my mother said. Which doesn't explain why my grandparents kept this antique, though maybe my own desire to have it now could explain it: this bottle is cool. It is gray glass, rectangular, and slightly smaller than a Zippo lighter, with a red label that says it was manufactured in Detroit. The black, rice-sized pills it still contains may or may not be what the label advertises, may or may not still be potent or poison. Never taken, but never thrown away. A time capsule of mysterious medicine.

////////////////////////////

In the last year, elderberry lozenges have appeared as a medicinal impulse-buy near the cash register of my local natural
grocer. They look like Life Savers. Their label promises to "support respiratory health" and "prevent flu." Their marketing
would have me believe I'm partaking in an ancient remedy.

And I am. But as far as I can find in my botanical-
medicine books, elderberry isn't mentioned as an antiviral
tonic. It is suggested as an eye soother, a diarrhea inducer,
a balm to asthma and epilepsy and dropsy. It gets the extra
fluid out and calms seizures if one makes a broth of the inner bark and drinks it every fifteen minutes. Nothing about
viruses (which weren't discovered yet, for the older books
like Gerard's and Culpeper's), and nothing about everyday
health maintenance—except the cooling effect the leaves
supposedly have, which, according to ancient Galenic ideas
about the four humors of the body, would bring a choleric
person back into balance. I can confirm that elderberries
were generally used as medicine for a very long time, but
the lozenge taker who assumes her cold medicine has been
passed down through the ages is partaking in a fantasy.

Elderberry lozenges are no more and no less effective
than Emergen-C or Airborne, those paradise-flavored vitamin powders that make similar promises and, though we
cannot say for sure they cure the common cold, can—like
anything presented as medicine—trigger a powerful and
observable improvement in health, especially if they're suggested to us by a trusted and caring source. Placebo studies
show that even when people consume pills from a bottle
labeled PLACEBO, they experience a beneficial effect. We

can know the medicine is fake and still be helped. The act of treatment, the ritual of taking care of ourselves, helps improve our well-being.

Current placebo studies attempt to understand what poets and children have always known: placebos work on a symbolic level, powerful for what they signify, not for what they chemically contribute. They are medicine the way kind words can be medicine, possessed of soothing properties because of how we relate to them.

I've taken some poetic license with "placebo" here. Current medical literature distinguishes between sugar pills, which are fake pills with real meaning, and words, which are not placebos but can promote positive effects through their symbolic power. I'm not a doctor, and I don't see a meaningful difference.

After foraging through my family's archives, I needed to harvest something more nurturing. The elder bush at the edge of my parents' land was not stooped but sort of looped, with branches that arched and swayed. I caught the tip of the nearest one, heavy with blue berries and purple stems that should not be eaten. All my sources say elderberry stems are poisonous. The raw berries contain toxic substances, too, and should be cooked before consumption. Cooking neutralizes the definitely poisonous glycosides and concentrates the possibly helpful anthocyanins. To pick them, I'd planted myself with one foot down a steep slope that dropped off toward a creek. Mom had a better idea: cut off the berry clusters at their apical stems, gather them in a bag, then remove the berries from their stems while sitting next to her at the kitchen counter.

These are silver-coated blue elderberries, and sour. My elder uncle recognizes them immediately, says he used to eat those as a kid in McCook, Nebraska, when he was out looking for trouble while my grandmother was at work. This was before his stepfather, my grandfather, moved to town for an internship at a pharmacy near the Woolworth's where he met my grandmother. My uncle doesn't remember elderberries as being blue. His were black. *Sambucus nigra*. These black elderberries are the fruit whose extract is all the rage right now.

According to Culpeper, elderberry "mollifies the hardness of the womb, and brings down the menses," which sounds to me like code for abortion, but might be what it says it is, a potion to regulate menstruation. Herbal medicine remembers what supplement labels don't report: its purgative effects are powerful. Use the roots to induce vomiting, the berries to move urine and stool. Those who aren't prepared for these properties will experience their medicine as poison.

Without speaking, insert a sprig of elderberry into the ground and fever will break, one folk remedy goes. To cure rheumatism, string a bag of elderberry pulp around the neck, says another. Eat the boiled root to help baldness. Or rub warts with elder leaves.

To copy an elderberry syrup my herbalist friend makes, I simmer elderberries in a 1:3 ratio of berries to water until the liquid is reduced by half, then remove them from the heat. Adding unfiltered local honey contributes a little unproven allergen therapy to elderberry's supposed antiviral properties, and a little romantic fussiness. Honey also makes the elderberry syrup too sweet to taste like medicine. "Let's add vodka," my younger uncle suggests. He's a family doctor. He doesn't believe I'm making anything more efficacious than simple syrup, but he's kind enough not to say so. And anyway, between vodka and elderberry decoction, vodka in small amounts has more proof as an antiviral.

But it is less loving.

Elderberry Syrup

In *The Master Book of Herbalism*, Paul Beyerl calls elderberry "a Queen among herbs" that once identified and eaten "is a plant always familiar as a friend." He advises we gather the berries and flowers in season, remembering that "those lovingly collected beneath a full moon are prime."

For millennia, herbals that catalogue the medicinal properties of elderberries and other plant-based remedies have been featured in the kitchen libraries of heads-of-household, homemakers, clergy, and medics as guides to curing common ailments. Herbals are also called "leechbooks" if you're a ninth-century Anglo-Saxon, "braucherai" if you're eighteenth-century Pennsylvania Dutch, and "DIY homemaking zines" if you're a twentieth- or twenty-first-century riot grrrl, to name a few variants. They teach the reader how to heal thyself and promote everyday well-being, and through their advice encourage a particular religious, philosophical, or political set of beliefs to play out within the basic tasks of living. A modern herbal or herbal-descended book might prescribe ginger and hot water instead of Mylanta to soothe a sour stomach, or baking soda, salt, and white vinegar to clear a drain instead of watershed-and-pipe-killing Drano. The result, besides empowering the reader's self-care, is a kind of literacy: we learn how to read the plants around us and use them to make our lives better.

Nigel Slater's *Ripe* (a cookbook/memoir/reference that reminds me of an herbal) and Paul Beyerl's *The Master Book of Herbalism* (a Wiccan- and astrology-influenced herbal) both refer to historical uses of elderberries and present their uses for modern audiences, pulling on some of the same traditional knowledge, recycling it, keeping folk wisdom and common knowledge alive. Slater's writing is more personal and detailed, grounded in the "I" and lyric in his descriptions, whereas Beyerl is concerned with the magickal and medical uses of the elderberry. "The leaves are considered cooling," he writes, in what I interpret as a flashback to Galenism's hot/cold, wet/dry humoral compass, a debunked system of thought that persists as a way to explain how to restore balance to a body that's lost it. The blood-type diet, for example, is basically humoral medicine glazed with a topcoat of science; it works because it promotes a healthier diet.

Neither author suggests the recipe I've come to know elderberries for, a decoction of fruit sweetened with honey that my herbalist and herbal-curious friends make to support immune-system health. My version now requires my parents' fruit, their protection and love enhancing their elderberries' already symbolic powers. Substitute your own fruit for theirs in any way that amplifies your feelings of being protected against sickness and harm.

Yield: varies

Elderberries (whatever amount you can gather)
3 times as much water
Honey (see procedure for amount)

Combine the elderberries and water in a Dutch oven or stockpot, and heat at a low simmer until half the liquid remains. Let the liquid cool long enough to become comfortably warm, strain, and then sweeten it with honey—as much as you wish. I use a 4:1 ratio of strained elderberry liquid to honey. Store the bottle in the refrigerator. Swallow 1 spoonful a day. Or 7. Whatever you need to stay well.

Elderflower Cordial

To test this cordial, I harvested flowers from my neighbors' landscaping—not a traditional black or blue elder, but a black lace elder. It's a beautiful bush, with white-pink blooms above purple-black leaves that look more like Japanese maple than wild elder. The flowers smell more like anise than black or blue elder-flowers, but the flavor and scent of the cordial are the same. The color isn't—elderflower cordial made from black lace elder is pink; expect a light-yellow color from more traditional preparations. All different kinds of elderflowers work for this recipe; just keep in mind that harvesting flowers now means sacrificing berries later.

To harvest elderflowers, cut them at their apical stems just above where all their smaller stems attach to one stem. Bring a small cookie sheet or flat-bottomed basket so they don't get crushed as you gather them. Pick flowers that have opened com-pletely, and try to pick before the hottest part of the day. Hot flowers won't have as strong a scent, and they'll be more stressed. Picked elderflowers wilt quickly, so use them immediately.

Elderflower cordials traditionally flavor gooseberry desserts, but I love them with peaches. Dress fresh peaches with a splash of elderflower cordial, add a couple tablespoons to a batch of peach preserves during the last minute of cooking, use as preserving syrup for canned peaches, or add a couple tablespoons to peach-pie filling before piling the fruit into the crust. Or fill a tumbler with ice and sparkling or still water, add a tablespoon of elderflower cordial, stir, and enjoy on a hot day. Use as a nonalcoholic substi-tute for St-Germain. Float a spoonful in champagne, or stir some into a gin and tonic for a subtle floral flavor.

Yield: about 8 cups

20 elderflower heads (just the large flowery umbels, not
    the thicker stems or leaves)

5 cups water

5 cups sugar

2 lemons, quartered

2 limes, quartered

¼ cup citric acid

Shake the flowers to dislodge any critters. You won't get them all, but get the big ones. Don't worry—all bugs will be strained out by the end of the recipe. Don't rinse the flowers; rinsing dislodges the elderblow, the part that makes them smell and taste so good.

Elder stems contain cyanogenic glycosides, just like cherry kernels do. This is easy to avoid, but, to be safe, trim off all thicker stems. I trim elderflowers over a large bowl and let the flowers drop into the bowl, then throw the de-flowered stems away.

In a large saucepan, make simple syrup by heating the water and sugar and stirring to dissolve. Remove from the heat after the sugar has dissolved.

Place the flowers, lemons, limes, and citric acid in a large bowl, then pour the hot simple syrup over all. Weigh down the flowers with the lemons and limes. Give the simple syrup a little stir to make sure the citric acid has dissolved, then let sit, covered, overnight.

The next day, strain the cordial through a jelly bag or a muslin-lined sieve. Let it drip slowly. Discard the solids. Bottle the cordial in clean bottles or jars, and store in the refrigerator. Some sources say it must be used within 6 weeks; I keep mine as long as it smells and looks fresh. One batch was delicious and drinkable for more than 2 years.

To preserve the cordial for later, freeze it in batches, leaving at least an inch of headspace in the jars so the cordial has room to expand as it freezes.

F: Faceclock

Taraxacum officinale
Asteraceae (aster, daisy, or sunflower) family
Also known as dandelion, blowball, puffball, cankerwort,
witch's gowan, milk witch, lion's tooth, monks-head,
priest's-crown, piss-a-bed, pee-a-bed, swine's snout,
wild endive, pissenlit

We have paid someone to poison the lawn so that we don't have to pull dandelions, but I'm pulling them anyway. It rained two days ago, mild warmth since, so the ground gives the flowers up without much fuss. I can't hear the dirt gasp, but I can feel it every time my trowel mines a perfect root. The applause of falling soil.

According to John Kallas, author of *Edible Wild Plants*, dandelion flowers open and close once a day for three days. On the fourth day, they open as seed heads, also called puffballs or faceclocks. He writes that, according to folklore, the number of times you have to blow on a faceclock to clear it of fluff tells you the time—or whatever seeds are left after three puffs. "Einstein would love this method," he writes. "Time would be relative to each puffball you picked up."

A common child's game calls for blowing out face-clocks like birthday candles, or using the thumbs to behead blooms while chanting, "Mama had a baby and its head popped off."

To weed dandelions, I do as my parents taught me: push a trowel into the soil at the crown and shimmy it until I feel something pop. I gather the dandelion where leaves meet stem, pull out the whole root, shake off the loose dirt, and throw it into the wheelbarrow.

This year, my spring depression once again inspires yard work. It looks like this: I'm talking on the phone to my oldest friend about how I don't know why the only thing that feels okay is gardening. It is her thirty-fifth birthday; I'm pulling dandelions. When I look over at my wheelbarrow, I'm surprised to see that all the yellow heads have disappeared. They've been replaced by cotton halos, emergency parachutes poised to catch a gust of wind and ruin all my work. I say, "Hold on, Lindsey," put the phone down, and turn the weeds in the wheelbarrow over, smothering the seed heads.

"What I'm hearing isn't that you love gardening," she says when I return. "What I'm hearing is, you love tapping into your life-source through creative work."

We share a therapist. I can hear Anne in the careful way Lindsey tells me what she thinks about my feelings, how she goes to the root, holds it up, and shakes it free. Plus that woo-woo word, "life-source," which Anne doesn't use (with me, anyway) but Lindsey wouldn't have used before Anne.

"Who's to say pulling dandelions doesn't tap into that same source?" Lindsey says.

Overwintered dandelions tend to bloom in synch, bursting from the lawn in April or May, depending on where you live. They like the cooler months and direct sunlight after partial shade. Growth in hot sunlight makes the leaves bitter, as does age. You can rip dandelions out by their leaves, but the root will regrow the plant the following season. You can prevent more dandelions by cutting off their flowers before they transform into faceclocks, but this will spur the plant to grow more faceclocks. Their yellow blossoms—which I keep calling "heads," as if they are people—make them seem cheerful. "King of village flowers!" Vachel Lindsay's poem names them. "Each day is coronation time," he writes. "You have no humble hours." In at least one instance, dandelions are king of the jungle, too: "I'm just a dandelion!" the Cowardly Lion sings with a swish of his wrist as Dorothy and the gang begin their long skip down the Yellow Brick Road.

Dandelions do have crowns, but those crowns are not their flowers. Their crowns are where their leaves radiate from the dirt. This is the part that can be gathered by hand into a mass of leaves strong enough to keep the plant in one piece as a gardener rips it up by the root.

The spring when we bought and moved into this house, the upheaval of our lives or the stress of it or something—maybe it was just spring—brought on a depression so deep in me that I didn't want to exist. Not suicide. I wanted to be inert, in the dark, asleep. This had happened before, but not so darkly. I stopped drinking. I exercised. I went to therapy. Sam tried to help, but he was tired, too, exhausted with moving. He needed my help. He suggested I might feel better if I helped him with a thing.

I tried this. I lifted boxes and arranged furniture. When he asked me where I wanted the couch, the painting, the bookshelf, it felt like being pried open. Something rude about it. I don't know, I said, because to say the truth (*I don't care*) would've been mean, and ultimately not true—I knew that when I woke up from wherever I was, I would have strong opinions about where everything should go. But right then choices felt physical, like being worried and poked and shoved. I don't know, baby, I told him. I need you to choose, I said.

"Why don't you try resting?" Anne had suggested.

I tried resting. I crawled into bed and imagined I was a bulb in the ground in winter. I imagined I was alive under dirt, a caterpillar, one of the spring-green coils of worm I unearth now with my trowel when I dig dandelions.

"If you like pulling dandelions," Anne suggests, "why not pull more dandelions?"

This year, like last year, I gather them weed by weed, digging up just the ones that have flowered, their yellow heads betraying their locations and helping me draw a line within my task: kill the obvious ones first. Young dandys pock the lawn, too, camouflaged by immaturity. Soon all but the last, coolest corner of the lawn is clear. Where I've been, the ground is bare and disturbed. It looks like I've cleared the lawn away, which means our suspicions were right: most of the lawn *was* dandelions.

It is satisfying, this scarring. Grass will grow over it, or moss, or more dandelions.

When I was a teenager, I considered being a nun. Never, not even once, did I consider being a housewife. Yet sometimes I look at my life, which I have lived as a certain kind of feminist and a certain kind of artist, and I see another woman who's struggling to make her peace with being at home.

While I grope for a tranquil state of mind, I spend the best hours of my writing day on yard work, whose main external reward is the good opinion of our neighbors.

If left to mature for two or three years, dandelion leaves grow large and tender, with roots as big and deep as their tops are tall and wide. To harvest the leaves, cut them off at the base. To eat them, develop a taste for bitterness or make a plan for dealing with it. The easiest and most delicious way is to boil the leaves, checking on the bitterness after four minutes and continuing to boil gently until they're tamed to a pleasant bite. Changes of water aren't necessary. Dandelion leaves can be boiled directly in soups as a substitute for other bitter greens; they give the soup a satisfying, earthy flavor. Yellow dandelion flowers, no matter how old the roots, are fresh each spring, edible and sweet. Their bracts—the green leaflike structures around the petals—are bitter. To make a salad with dandelion flowers, remove the petals from the bracts. To harvest what Kallas calls the dandelion's heart (the stemmy part between the roots and leaves also known as the root crown), catch all the dandelion's leaves in one hand and yank them upward a bit to loosen the heart from the dirt. Then cut through the dandelion at the dirt line, leaving the stunned root to grow another crown. Remove the leaves, trim any stems to an inch long or shorter, then soak the hearts in water to loosen any dirt. Sauté with butter and garlic, tasting for bitterness before removing the hearts from the heat. If they're still too bitter, keep cooking. Yard dandelions are less delicious than field dandelions, or so I'm informed when I tell my across-the-alley neighbors that this year I might consider our weeds a crop.

"You'll kill all the critters in your soil!" Josie, our across-the-alley neighbor, scolds. We shouldn't have told her about the poison. She's trying not to show how upset she is. She is failing. "Dandelions attract pollinators!" she says.

And they do. I see a bee visiting the yellow heads in my wheelbarrow. Bees in the centers of the dandelions I haven't yet dug.

"If you want, I'll take care of your lawn," our next-door neighbor offers. Earl's property line is lush and green. Ours is scabby, mossy, full of weeds. After dinner, he sits on his porch, smoking and drinking, watching our weeds threaten his lawn. He's lived in that house for thirty years. Hasn't retired from the golf course yet, but will soon. We won't let him take care of our grass, but we're putting weed killer on our lawn for him. To be good neighbors.

///////////////////////

One weeknight, around eleven, Earl stands in the picture window that faces our house, his lights ablaze, stark naked. I tell Sam to get up and leave and then walk back into the room so he can see Earl without looking like he's trying to see Earl. Sam says he will not. He says he'll just have to trust me.

Doesn't Earl see us here, in our own lit living room, eating pizza on the couch? When I sneak a glance, Earl's bending over. He puts one leg, then the other into a pair of shorts he's drawn from a laundry pile on the table. I look away before he can raise his head. The lights go out.

We cancel the poison.

We say we're canceling for Josie and Liz, transplants from Chicago by way of Seattle, organic gardeners who spend their first spring in their new house tearing out the grass the previous owner had Miracle-Gro'd into obscene health. They put in apple and filbert trees, currants and gooseberries, blueberry and goji and aronia bushes. For them we ask the lawn service to fertilize, aerate, and seed the lawn—no herbicides.

When the service comes, I ask the guy for advice on how to get rid of our dandelions organically. He says, "I don't know, ma'am. I'd have to do some research."

"So you have no advice for me?" I ask.

"Nope," he says.

I think he enjoys saying "nope."

He straps a plastic dispenser to his chest, cranks a wheel that spits fertilizer into the grass, and charges us fifty bucks.

"Why even have a lawn?" Lindsey asks. Josie and Liz asked this, too. If we don't have a lawn, we have to have something else, I say. We can't just have dirt.

It is spring, and the foragers of Instagram are collecting yellow dandelion heads to make plant-based dyes. I see recipes for dandelion wine, dandelion vinegar, dandelion shrub. How do you make a universal plant exotic again? Dye your own clothes with flowers, harvest your own roots for coffee, draw baths to soak the bite from foraged leaves.

These look like ways to make a sport out of weeding, not like anything actually delicious. They are not chores or duties, they are choices, but they take our time and wring it down the sink just the same. Capturing the occasion for social media gives the task performative utility: I made something today. I was productive. Look.

When I was younger, before I ever dreamed of owning a house, the idea of loving to garden, of spending my time obsessed with a small plot of safe land, sounded so depressing. Women humming in their gardens like bees trapped in a jar. It looked like the work of adults consoling themselves for their boring lives.

"They're the opposite of humans," Sam says when I tell him how dandelions can form faceclocks even after they've been pulled. "Our reproductive system shuts down first, protecting the self. Dandelions use the last of their energy for the next generation."

Sam worries that if we have a child my depression will overwhelm me—and our relationship, and our family. "I'm not crazy," I say, and he says, "No, no, baby, that's not what I mean."

Before Sam, I lived with W, a man whose moods so overwhelmed my own, I stopped noticing I had them. After that I lived alone, with no partner to protect from my moods—or inflict them on. My storms were my own to ignore or surrender to. Now, steady with Sam—a homeowner even!—I am the partner whose health dominates the weather of the house. Sometimes I can warn Sam of my mood—I call it a mental flu. Sometimes he has to warn me. Sometimes I can climb out of it just by telling him it's happening, and sometimes it feels better to be silent, to sink in, a conspiracy with myself that gives me bitter pleasure in excluding him.

Dandelions were brought to the United States for medicinal purposes by the Pilgrims—or so goes the legend. Native to Europe, they naturalized in North America after European colonization and have since spread all over the world.

The consumption of dandelions is not a recommended treatment for depression; rather, they're a diuretic, which explains "pee-a-bed" and "pissenlit"—other folk names for the flower.

There *is* evidence that gardening makes us happier. In a letter to her friend Maria Whitney, Emily Dickinson writes, "Intrusiveness of flowers is brooked even by troubled hearts. / They enter and then knock—then chide their ruthless sweetness, and then remain forgiven . . ."

///////////////////////

I want to try a dandelion-coffee recipe Anne was telling me about, the detoxifier, so I start saving the roots. You roast them in the oven, she says, nothing special, then grind them up in a coffee grinder and use them instead of coffee.

I harvest almost the entire lawn until I get to the back corner, the coolest nook of the backyard. This is where I discover that the neighbor's cats have shit in the centers of the biggest, lushest dandelion crowns, a game of lawn darts the spring rain hasn't yet reset.

That's enough. I am sick of dandelions.

I leave the remaining weeds and the shit where they are and throw the whole wheelbarrow of roots into the trash.

I am not opposed to eating my weeds, but that's not why I go to the backyard on my hands and knees. I don't want the Instagram photo op or the piety of the forager. I want the act of extracting them, the tug and release and the release within me, my rotten brain, dirt falling away perfectly when the dandelion is young and the earth is wet, so it feels like popping a bubble, or like the long, straight cut of a knife in rhubarb. A perfect motion. The plant remade. I am clearing the grass of weeds so there can be more grass or new weeds, or until something within me clears, too.

I don't want the grass. I want, only, to have made a choice about the grass.

Faceclock Coffee

Pulling almost all my backyard dandelions eradicated the following year's crop. When I went to gather more to attempt faceclock coffee, I didn't have enough. Across the street, there's a house that's been empty for longer than we've lived here, gutted and supposedly unlivable. From the outside, all we can see is an attractive front porch and a yard crowded with dandelions. A bonanza of unsprayed roots, I assumed, figuring that if the house flipper who owns it can't even get the water running, when would they have sprayed for weeds? These dandelion crowns were huge, but their roots were fairly small—only 3 to 5 inches, depending on the root. I gathered a 7-gallon bucketful of dandelions and my yield of roots was less than 2 cups.

In *Edible Wild Plants*, John Kallas admits he's not very familiar with dandelion roots, does not know for sure if new roots are tender and old roots are not, and mentions eating them, but not drinking them as coffee. In *Stalking the Wild Asparagus*, Euell Gibbons mentions dandelion coffee but does not provide a recipe for it. Kay Young's *Wild Seasons: Gathering and Cooking Wild Plants of the Great Plains* suggests a 350°F oven, and says, "For each cup of beverage you will need to dig one medium-sized root." Not knowing whether her medium is the same size as my medium, I followed instead the proportions I like for pour-over coffee.

Springtime tonics are traditionally drunk to invigorate the body, as if we, too, had sap that could run high or twigs that needed pruning. This coffee (which is also commonly referred to as a tea) dries the mouth slightly, the way coffee does, an indication that it's a diuretic, a sensation that could be blunted with milk if you wish. It's a purgative, in other words, that brings the body back into balance in the Galenic sense, stimulating the liver and kidneys according to herbal lore—which is a nice way of saying that, as with coffee, soon after you drink this brew you'll need to use the restroom.

Yield: varies

Dandelions, roots intact (as many as you can find, or
    have the patience to pull)
Vanilla extract (optional)
Sugar (optional)
Milk (optional)

To prepare the roots, snip them from the root crown and place them in a salad spinner, discarding the rest of the plant. Run warm water over all the roots and let them sit for a bit, agitating and rubbing them every now and then to encourage the dirt to wash off. Drain, discard the muddy water, then run another bath and repeat until reasonably clean. Finally, use a vegetable brush to remove the rest of the dirt gently, discarding any roots that are soft or buggy. Do not peel, and do not remove the root hairs.

Preheat the oven to 350°F. Roast on a baking sheet for an hour. Then do as Young says: "Check for doneness by breaking a root in two—when done, the root should be very dry and brittle and smell a little like chocolate. Small roots should be brown most of the way through, but larger roots will have a lighter-colored core." I can't confirm the chocolate smell, but they do have a nice toasted-potato-skin aroma. Young advises to keep an eye on the baking. Too little and the roots will be tasteless, she says. Too much and they'll taste burned.

After 1 hour, check on the roots. Pick out the smaller ones first. If they snap easily in two, remove them from the baking sheet and set them aside. Some roots have thick and thin parts, which will be done at different times. I snap the thinner, fully roasted ends off and put the softer, thicker parts in for another 15 minutes. Then repeat the process of checking to see if the roots snap easily. If some don't, roast them a little longer, checking on them every 5 minutes. Allow all the roots to cool. Once they're cool, store them in a covered tin and grind them just before drinking.

To prepare faceclock coffee, grind the roots roughly in an electric coffee grinder or spice grinder (Young recommends a mortar and pestle, which I'd usually be enthusiastic about, but in this case it makes the task more difficult). You'll make a chunky powder that smells like plaster and creates a fine, plastery dust cloud as you open the coffee grinder. To brew, scoop 2 tablespoons ground dandelion root into a Chemex strainer and pour 2 cups boiling water over it. For the uninitiated, the dominant flavor is a vegetal bitterness that lingers. Cut the bitterness with a single drop of vanilla extract, a dash of sugar, and a splash of milk, if you like.

Faceclock Greens, Fennel Sausage, and Barley Soup

This is an early-spring dinner for days when there's sun enough to harvest dandelions but not enough to leave your sweater inside. Scissor only new, tender leaves from the plant, preferably before hotter weather increases their bitterness. Even when tender, dandelion leaves will be quite bitter (one of dandelion's nicknames is "wild endive," a reference to its bitter distant cousin in the aster family). That taste will be stewed away in the soup, leaving a nutrient-rich leafy green soaked with an earthy, savory broth, righteously delicious like kale or bok choi.

I make the pork bone broth this recipe calls for by roasting pork bones (enough to fill a half-sheet baking pan) at 400°F for 20 to 30 minutes, until they're spitting and blackened and sitting in a shallow pool of rendered fat. Discard the fat, place the bones in a stockpot, add a halved onion, two bay leaves, eight peppercorns, two or three chunked celery ribs, two or three chunked carrots, some garlic cloves, and 2 tablespoons apple-cider vinegar. Cover everything completely in water. Bring the stock to a boil, then reduce the heat until it barely bubbles. Leave it that way, covered, overnight, until the bones are porous and brittle and the broth is rich and meaty. Then strain the broth and discard the solids. Refrigerate the broth until it has completely cooled, then remove and discard the fat that solidifies at the surface. Use the broth in this soup.

Homemade tomato paste adds more depth and acidity to the broth. Bright (not bitter) dandelion petals and a parsley-caper gremolata finish the soup. The fennel sausage and gremolata are elements of recipes from *A Platter of Figs and Other Recipes,* by David Tanis, repurposed here in this reinvention-by-substitution of an Italian-style kale, bean, and sausage soup.

Yield: 8 servings

FOR THE SOUP

1 pound ground pork

1 teaspoon whole fennel seeds, crushed in a mortar and pestle

½ teaspoon crushed red pepper

2 garlic cloves, minced

1 teaspoon salt, plus more to taste

1 pound fresh saucing tomatoes (Roma, San Marzano, or other sauce variety), cut into 1-inch dice, or one 14-ounce can diced tomatoes (with liquid), or a couple heaping tablespoons of tomato paste

2 tablespoons olive oil, divided

1 onion, cut into ½-inch dice

1 cup barley

12 cups pork bone broth (see headnote)

1 bay leaf

2 sage sprigs

10 to 16 ounces dandelion greens, tough ribs removed, washed and roughly chopped

Black pepper, to taste

FOR THE GREMOLATA (OPTIONAL)

1 bunch Italian parsley, chopped

Zest of 1 lemon, finely grated or minced

2 tablespoons chopped capers

Pinch of salt

Dandelion flowers, bracts removed and discarded, petals loose (optional)

First, make the sausage—perhaps while making the bone broth, since both will benefit from being made a day ahead. Combine the ground pork, fennel, red pepper, garlic, and salt, working it quickly by spoon or hand. Cover it, and store in the refrigerator overnight.

The next day, make a tomato paste by stewing the tomatoes in a small heavy-bottomed saucepan over medium heat. They'll be very juicy at first, but as they stew, the juice will evaporate. Stir occasionally to prevent sticking. Cook down until the tomatoes are a thick paste, only slightly watery.

While the tomatoes cook, continue with the rest of the recipe.

In a heavy Dutch oven over medium-high heat, fry the sausage in 1 tablespoon olive oil, breaking it up into smallish chunks, until it has fully cooked and slightly browned. Lay the sausage on a paper-towel-lined plate and set aside.

In the Dutch oven, remove extra fat, or add more olive oil, to reach about 1 tablespoon of fat total, then sauté the onion in that fat over medium heat, adding the tomato "paste," until the onion is soft and a little brown, scraping up sausage bits from the bottom of the pan. If those bits are stubborn, add a little water or stock and scrape again. Stir in the barley, coating the grains with oil and onion and letting them sauté for 1 minute. Add the bone broth, bay leaf, and sage. Increase the heat to bring the broth to a gentle boil, then cook, covered, for 10 minutes.

Add the dandelion leaves and sausage to the pot, and reduce the heat so the soup stays at just under a simmer. Taste, and add salt and pepper as needed.

Cover the soup, and let it sit on the stove at just under boiling for an hour or two, or until dinnertime, to let the barley absorb more broth and the dandelions' bitterness subside. It is, like most soups, best on the second day.

Before serving, remove the sage and bay leaf. For extra zip to finish, combine parsley, lemon zest, capers, and a pinch of salt, and sprinkle this gremolata over each steaming bowl of soup. Then sprinkle dandelion petals over all, if you like. Serve with crusty bread or biscuits and butter.

G: Gooseberry

Ribes uva-crispa (European gooseberry) and
Ribes hirtellum (North American gooseberry)
Grossulariaceae (currant) family
Also known as goosegogs, feverberry, carberry,
deberries, krusbaar, groser, grozet, honeyblobs,
feaberry, squinancy berry, the Scottish hairy grape

Gooseberries are sour. They are sour even when ripe. Not punishingly sour like cranberries can be, though they are similarly sized, a small roundness that suggests they would be pleasing to eat one by one. Nor do gooseberries play at sour the way green grapes, those utility players of the lunch sack, play at sour. Gooseberries are sour like you've arrived before they were ready for company, like they wanted you to see them in a better dress.

Where I come from, gooseberries are a rare find and carry with them a mythology from elsewhere, a vague sense that they are a fruit One Should Know, perhaps from Literature.

I love them because no one else of my acquaintance loves them, and to be without competition delights me. I love to bring a pint of ripe gooseberries to cocktails with people who've never seen gooseberries before, so that now when they hear the word "gooseberry" these people think of me. I can bring gooseberries to cocktails in July only, during a yearly conference on Washington's Olympic Peninsula that coincides with the two or three weeks when Jefferson County's small gooseberry crop ripens. I'll have picked the gooseberries myself that day, following the gooseberry farmer's instructions to grab the tip of the branch and lift it up as far as it will go, exposing the fruit and thorns. Every year, he tells me to pick clean. Any bush I leave partially harvested will make more work for him, he says. Later that day, when I eat the gooseberries with friends, we will sit outside, because from here we can see Puget Sound, whose water in daylight is so brilliant it shows the flaws in my own corneas, black tracers over clear, hard blue. We'll eat these gooseberries and drink too much whiskey or white wine to finish the poem or essay or story we wanted to finish the next day, but not so much we embarrass ourselves. No, gooseberries are sour like Campari is bitter, like they're possible to enjoy without embellishment or dilution if one has a taste for them, but the rest of us must make an effort.

To friends I have described gooseberries as having a vegetal undertaste like a tomatillo. This is true only for green gooseberries. *Cape* gooseberries are related to tomatillos and, like tomatillos, clothe their fruit in delicate papery hoods, but Cape gooseberries are not true gooseberries. True gooseberries are purple or maroon or almost black or pale green, and they are hoodless. They look like giant currants, to which they're closely related, with dark lines that spool from calyx to stem, taut and jagged like stretch marks.

Chekhov does not tell us the color of the gooseberries in "Gooseberries," the second of a trilogy of stories he finished in 1898. "Gooseberries" is a tale of two friends, Ivan and Burkin, who visit their friend Alyohin's gentleman farm. Ivan tells the men a story about his brother, Nikolay, a clerk who dreamed of owning a country estate with a gooseberry patch. Nikolay marries a homely but rich widow, then starves her to death with his penny-pinching. The land he buys with her money is hot and badly landscaped, with a stream polluted by a glue factory and no gooseberry bushes. His servants look like pigs. Even Nikolay looks piggish, fattened on his obsession, planting gooseberry bushes and suing the glue factory like the entitled man he strove to be.

When Ivan visits, Nikolay serves him the first harvest of his beloved gooseberries. To Ivan, they are hard and sour; to Nikolay, delicious beyond reason. All night, he sneaks out of bed to eat more of them, keeping Ivan awake. Ivan can't understand why Nikolay would enjoy such hard, sour gooseberries, any more than he can understand why Nikolay would give up the struggle of city life to bloat himself on the pastoral leisure of the landed class. Happy people are horrible, Ivan thinks. All happy people should be followed by a little man with a hammer to smack them when they are happy so they cannot forget the suffering of others.

Elsewhere in the nineteenth century, English backyard gardeners were in the midst of a competitive frenzy of gooseberry growing, as Lee Reich describes in *Uncommon Fruits for Every Garden*. "Whereas wild gooseberries weigh only about a quarter of an ounce, a gooseberry show winner in 1817 weighed in at one and a quarter ounces; by 1852, a winning fruit tipped the scales at just under two," catching the eye of Charles Darwin as he composed his arguments for natural selection. Unusual growing strategies included allowing chickweed to colonize the gooseberry bed, supposedly promoting a cool air for the growing fruits, and, "perhaps the strangest practice," resting a saucer of water below the gooseberry so its calyx rested in the pool, as if the berry could "suckle" water through that nipplelike berry part.

Across the Atlantic Ocean, European gooseberries were little loved by American palates and didn't thrive in American soil; hybrids of native and European gooseberries produced a plant that carried white pine blister rust, fatal to pines but mostly harmless to gooseberries. Federal law halted the spread of hybrid gooseberries in the 1920s to protect American pine forests, which, today, are beset by pine beetles instead.

The gooseberry pies I encountered during the years I judged pie at the Iowa State Fair had green, almost gray fillings, made from a different sort of true gooseberry, one that I haven't seen fresh, though I'm told they can be found by streambeds in Iowa and Nebraska. Green gooseberries are often photographed with white elderflowers, because they ripen at the same time—early summer. This may be true in England, where gooseberries have an assured place in traditional cuisine, but in my experience they're not synched with elderflowers. I make elderflower cordial in early June, and gooseberry cheese in mid-July. Green gooseberries are photographed as gray-green mush layered with whipped cream, a dessert called a "fool" because, some say, anyone can make it, though Jane Grigson, in her *Fruit Book,* says "fool" has even simpler origins, that it's a word like "trifle" and "whim-wham," "names of delightful nonsensical bits of folly" that are "outside the typical range of the cookery repertoire."

Gooseberries have a powerful amount of pectin, the naturally occurring starch in fruits that transforms into sugar as fruit ripens, and, when heated, binds with sugar and acid to make the gel that sets jam. Hard, unripe fruit has more pectin than soft, ripe fruit. Ripe blueberries, blackberries, and strawberries have very little pectin; ripe quince, cranberries, and gooseberries have a ton. Knowing how to work with the pectin content of fruits takes pies and preserves to the next level, because the baker and jam maker skilled in this minutia of the preserving arts is able to adjust his or her recipe to produce a texture that's both tasty and tidy—not running off the spoon like soup or gummed up like Jell-O.

Jellies, jams, and preserves distinguish themselves from each other by how they treat fruit. Jelly is basically a fruit-pectin stock fortified with sugar, whereas jam is the fruit itself fortified with sugar. Jelly retains only the essence of fruit; jam includes the essence, appearance, and texture of it. Preserves can be confused for jam without consequence or anyone's even noticing, but technically they're a fruit-and-sugar combo cooked in such a way that the fruit maintains its shape. Preserves aren't as spreadable. They are the embalmment of fruit. "Preserves" is also the general term for jamlike and jellylike substances. Conserves are preserves with nuts in them.

To set, all jellies, jams, preserves, and conserves require a certain ratio of pectin to sugar to fruit. One must boil off a certain amount of water to concentrate the sugar, which allows the preserve to reach a temperature of 221°F—just nine degrees higher than the boiling point of filtered water, and slightly lower than a candy chart cares to describe. Hotter than that or boiled too long, the sugar hardens into a pliable but stiff substance, in between the food categories (not jam, not candy, not spreadable or spoonable or cuttable or able to hold a precise shape), so that it remains uneaten not because it is not delicious (it's sugar and fruit!) but because one can't identify the culinary use to which it can be applied. Pies are cousins to fruit preserves in that they involve many of the same principles and players, but wrap their sugar and fruit in pastry that wilts after three days. The filling for a gooseberry pie needs very little help to thicken—maybe two tablespoons of instant tapioca, or two tablespoons of flour and two of butter. The best texture for pie filling and jam is one that can stand up on its own in a mound on the spoon, or within a double crust on a plate, but slumps a little, soft enough to spread on toast or cut with the side of a fork, an ideal state of being that's both sturdy and pliant.

In the gooseberry division of the most popular pie contest at the Iowa State Fair, there were never more than five pies, usually only two or three, and no one ever fought me for the opportunity to judge them. They were, as you may by now imagine, sour. Sometimes, after I finished judging, something like this would happen: I'd be congratulating the baker of the first-place gooseberry pie, and a woman with chemo-shortened hair would stand behind her, waiting to approach and say, "I'm Diane! We're family! Your mother's cousin on her father's side!" And we would laugh and hug and part and never see each other again. Or the prizewinning pie lady of Audubon, Iowa, would be direct this time when I asked for a photo, tell me why she wanted to say no. A tractor crushed her face years before, and it healed crooked. Couldn't see the scars in person, but on camera they were unmistakable. Her sun-furrowed face was friendly and open, but that's hard to detect in our selfie. Or I'd be wandering around the food hall on the last few days of the fair, inspecting the display cases where the first-prize preserves and desserts were housed, the layer cakes shaped like golf clubs and castles, the jars of pickled meat, the lofty pies whose meringue had broken out in beads of syrup, the jellies and jams and conserves as neat and peaceful as bricks. I'd see the grand-prize cupcake sporting a thatch of mold so subtle it hadn't yet been caught and escorted from the hall, and I would privately salute it, that rot, the force that inspired this competitive cooking, all our blue-ribbon efforts to showcase the harvest while arresting decay. To be in the company of people who enjoy "putting up" is, for me, to be with family. No one asked for the secret to prizewinning gooseberry pie. It's sugar, of course. And a judge who likes gooseberries.

Gooseberry Cheese

This recipe was inspired by a gooseberry cheese Jane Grigson mentions in her wonderful *Fruit Book* but does not provide a recipe for. I encountered a possible blueprint in Kevin West's (also wonderful) *Saving the Season*; his recipe for membrillo transforms the quince pulp left over from making jelly into a preserve of fruit so thick, it's served in slices like cheese. "Fruit pastes are one of the oldest confections," writes Jane Levi in *The Oxford Companion to Sugar and Sweets*, used "as medicines, travel snacks, sweets, and desserts, and as accompaniments to both sweet and savory foods." Fresh gooseberries and gooseberry jelly were recommended as a fever cure in sixteenth-century Europe, which is how it got the nicknames "feverberry" and "squinancy berry."

Here I substitute gooseberries for quince and do away with West's jelly step to hoard all the pectin and gooseberry flavor. Recipes fail all the time, especially when I'm first trying to write one, but gooseberry cheese surprised me by coming together easily. It has a sweet-tart-tannic, very *gooseberry* flavor—so satisfying, even plain from the jar. It won't be quite as stiff as quince cheese (you will not be able to preserve it in blocks, for example). Rather, it will be firm enough to cut from the jar and serve in slices, but soft enough to spread with a butter knife. If you're able to source immature gooseberries alongside mature ones, do so—their pectin will help the gooseberry cheese set up. Eat with roasted or preserved meats. A good goat's or sheep's milk cheese is a tasty pairing, too.

Yield: about 40 ounces

1 kilogram (about 6 cups) gooseberries (black or red
   or green, whatever you have—include some unripe
   gooseberries if possible)

¼ cup water

4 cups sugar

Juice of ½ lemon

Combine the gooseberries and water in a preserving pan. Over medium heat, bring to a soft boil, stirring occasionally, and cook until the gooseberries are soft and bursting.

Add the sugar and lemon juice. Do not be tempted to add these ingredients before the gooseberries have collapsed; adding sugar too soon toughens fruit skins and interferes with how the fruit releases pectin.

Bring the mixture to a boil over slightly hotter than medium heat, stir to dissolve the sugar, and continue to cook the pulp, stirring occasionally to prevent scorching. Reducing and thickening the gooseberries will take time—at least 45 minutes. The mixture should bubble busily, but not spit or foam. If it does, reduce the heat. The gooseberries will darken to a deep maroon, even if you're using green gooseberries. Once you can see the bottom of the pan when you stir the thickened mixture, and the gooseberries are silky and viscous, remove the pan from the heat and set aside.

While the gooseberries stew, prepare a deep canning pot with boiling water—enough to cover six 8-ounce jars. I add a little white vinegar, to mitigate the powdery white residue my hard water leaves on my jars after I boil them. Sterilize the jars by keeping them immersed in the boiling water for 10 minutes, then set them on a clean towel to cool and dry. Boil the lids for 10 minutes, then place them on a clean, lint-free towel, seal-side up, to dry. Rinse the bands and set them aside. Keep heating the pot of water on low until the next step of the recipe, when you'll use this water bath to process the filled jars.

Ladle the hot gooseberry cheese into the dry jars, leaving ½ inch of headspace. Wipe the rims clean, place the lids, and screw the bands on, fingertip-tight (tight, but not so tight you need strong hands to reopen them). Bring the water bath back to a boil and process the jars, keeping them immersed for 10 minutes.

Remove the jars from the water, and set them on the counter to cool.

Store in a dark cupboard at room temperature. Best eaten within 12 months. Refrigerate after opening.

Gooseberry Elderflower Frozen Fool

Gooseberry fool, a traditional English dessert of sweetened cooked gooseberry purée layered with sweet whipped cream, is by definition easy. This recipe complicates it. Starting with an American-style elderflower-flavored cream base (which is easier to make and closer to the ingredients of a fool than a French-style custard base would be), I add gooseberry purée and elderflower cordial to make a gooseberry dessert that requires only a little extra effort to enjoy.

If you haven't yet made the nonalcoholic elderflower cordial from page 65, try substituting elderflower liqueur. Alcohol has a lower freezing temperature than water and will create a softer scoop. Start this recipe the day before you want to eat it.

Yield: about 5 cups

2 cups heavy cream
1 cup whole milk
¼ cup elderflower cordial (page 65)
½ cup plus 2 tablespoons sugar
About 2 cups (350 grams) sour gooseberries
2 tablespoons water

Start by making the cream base. In a medium bowl, combine the heavy cream, milk, and elderflower cordial. Whisk in the ½ cup sugar until it's dissolved. Leave overnight in the refrigerator to get the cream really cold and meld the flavors.

In a small saucepan, cook the gooseberries, 2 tablespoons sugar, and water over medium heat until the gooseberries burst, about 10 minutes. Mash the fruit with a muddler. Chill completely in the refrigerator overnight.

Put a metal loaf pan in the freezer to use the next day. If your ice-cream maker has a component that needs to be frozen, make sure that part is in the freezer, too.

The next day, churn the ice cream according to your ice-cream maker's instructions. Churn the elderflower base and the gooseberry compote together, adding the cream, then the fruit, to the ice-cream maker.

Spoon the churned ice cream into the chilled loaf pan. Cover tightly, and freeze for at least 3 hours. Best served within the week, but keeps longer if you don't mind a little crystallization.

H: Huckleberry

Vaccinium ovatum (evergreen huckleberry) and
Vaccinium membranaceum (mountain huckleberry)
Ericaceae (heath) family
Also known as winter huckleberry, thinleaf huckleberry,
black huckleberry

As far as huckleberries were concerned, "the difficulties of domestication will be no greater than with the blueberry," wrote the horticulturalist U. P. Hedrick in his 1922 *Cyclopedia of Hardy Fruits.* He was hopeful. I like that in a scientist. But about the huckleberry, he was wrong.

Huckleberries refuse to be domesticated. To thrive, they need a particular spot within a particular type of forest. They need fires to consume light-hogging conifers and to muffle the soil with ash. Despite a century's efforts to breed bushes that thrive in monocultures at lower elevations, huckleberries still can't be cultivated as a profitable commercial crop. They must be gathered by hand in the woods.

For some Indigenous peoples and their descendants, picking these dark-purple, powerfully flavored berries— imagine a wild blueberry, but more sweet, more sour, more wild—is part of the social fabric of the tribe. It is not resource extraction. Nor is it agriculture as non-Natives generally understand agriculture. This reflects some biological truths that attempts at domestication have tried (and must try, by domestication's definition) to breed away: Huckleberries are wild. Their interdependence within their environment keeps them wild.

The English love a good lawn, or at least Captain George Vancouver did. When HMS *Discovery* entered Puget Sound in the late spring of 1792, he was amazed by the meadows he saw just beyond the beach. "Nature had here provided the well-stocked park," Vancouver wrote in his three-volume account of the voyage, "and wanted only the assistance of art to constitute that desirable assemblage of surface, which is so much sought in other countries, and only to be acquired by an immoderate expence in manual labour."

These "natural" parks were not natural at all, but the result of controlled burns by native tribes from British Columbia to the Willamette Valley, east through the Methow Valley, all the way to Montana. "Firing in the camas beds, huckleberry fields, oak groves, and tule flats, as well as other environments, took place *after* harvest, as a kind of post-use cleanup process, with ecological consequences in following seasons," writes Robert Boyd, the editor of *Indians,*

Fire and the Land in the Pacific Northwest. He details how
the tribes knew that fire could reset a landscape, turning
back the succession of plants to softer-fleshed annuals with
easily accessible nutrients (like berries) instead of longer-
growing, tougher perennials (like conifers, the "weeds" of
the huckleberry patch, as one of the Methow elders Boyd
interviewed considered them). They knew that fire adds
nutrients to the earth, creating ash-cap soils that huckleberries love while clearing the landscape of shady canopy. Fire one year meant a better harvest the next. Where
Vancouver thought he saw the hand of nature, he was
actually seeing strategic and careful resource management
by many different tribes.

Today's rules of huckleberry picking in the Inland Northwest are the social kind, restricting behavior to benefit
the community rather than the individual. LaRae Wiley,
a member of the Colville Confederated Tribes and cofounder of the Salish School, put it to me this way: "The
idea that you come in and you over-pick, you don't share—
that affects everybody. When you're in the huckleberry
fields, you're there with all the other animals. What you do
matters."

LaRae founded the Salish School with her husband,
Chris, in 2010. They teach Salish kids how to speak their
native tongue through immersion learning, with the hope
that children, who are natural language-learners, can bring
Salish back to their homes and help restore it to the community. She was in her early thirties when she first heard
Salish, at the funeral of an elder. "Growing up where I did,"
she says, referring to Cheney, Washington, "I didn't learn

much about my family's traditional ways. It's like my roots there were broken off."

Huckleberries grow below their leaves, so finding the berries is a matter of perspective. I like to bend down on one knee, as if I'm trying to look a child in the eye. As my center of gravity descends toward the earth, what had appeared to be a scrubby little bush suddenly reveals its prize.

You can rake huckleberries instead of picking them. One instrument for that looks like a covered hand-shovel with teeth. We're meant to swing it at the branches and let the teeth strip the berries off. This method strips leaves and branches, too, which robs next year's pickers of a good harvest.

Another fast-collecting method is to spread a blanket under the branches and shake the bush until its berries fall. This, too, shocks and injures the plant.

LaRae and her family never rake. They pick. It's slow work, but only if you're thinking about work in the short term.

The simplest way for me to pick huckleberries is to wait for August, point my car at the nearest mountain, and stop wherever the roadside looks not too well traveled and sufficiently bushy, promising a large harvest without grime from passing cars. These are the berries LaRae would leave behind for her elders to pick.

If you're younger, LaRae advises, go up higher, into difficult terrain, leaving the flat places and roadsides for elders. Give thanks. Some people leave tobacco before they start picking, some say prayers. Never pick all the berries on a bush. Leave some behind for the bears—they are our family, too. Pick, don't rake.

//////////////////

Since white settlement, but especially since the Great Depression, when so many people were so poor, huckleberry picking has been a subsistence option for all sorts of people who want piecemeal summer work. That's still true.

By the early twentieth century, Native huckleberry camps and white huckleberry camps were sharing the same forests. In the Indian camps, people relaxed before salmon season, spent time with friends and family and neighboring tribal members, gambled and drank, flirted and sang. Sometimes marriages were performed in the fields. In *People of the Dalles: The Indians of Wascopam Mission*, Robert Boyd records the frustration of one early Methodist missionary, Henry Brewer, with these camps. On a trip to the Mount Adams "whortleberry" fields (as Brewer calls the berries he encounters) in September of 1845, Brewer laments, "The absence of our Indian converts so long a time during the berry season, being surrounded as they are by every possible bad example, and separated from the watchful care of their teachers, in many cases proves very injurious to their piety."

In the white camps of the early twentieth century, extended families also picked and partied, but at the end of the summer they sold much of their crop. As food-preservation technology improved, people brought canning tools into the fields and processed the berries around the fire, moving camp every week or so to follow the berries up the mountain as they ripened.

Today, commercial foragers fill their buckets by picking and raking; there are no laws to restrict method. The harvest is never certain, and the labor is intense—two variables guaranteed to turn a food that's free for the taking into an expensive delicacy. A gallon of huckleberries that was $40 one year might be $90 the next.

On Washington State's public lands, pickers may take one gallon of berries a day, up to three gallons of berries per year. These limits are mostly self-monitored. To pick more than that, obtain a permit from the ranger. At the Pine Creek Information Center on Mount Saint Helens, you can pick up huckleberry maps in English, Spanish, and Russian, but not Sahaptin, the family of languages once spoken by the tribes in that area, and today nearly—but not quite—lost.

From what I've found in old horticultural texts and pre-twentieth-century literature, English speakers seem to give "huckleberry" as a common name to any wild, dark-colored, strongly flavored berry, regardless of genus. Like wild blueberries and cultivated blueberries, what I call huckleberries belong to the *Vaccinium* genus and the heath family. Huckleberries native to the East Coast, however, are usually not *Vaccinium* but *Gaylussacia*, a relative of blueberries and another member of the heath family. In *Cyclopedia of Hardy Fruits*, U. P. Hedrick's huckleberries were exclusively *Gaylussacia* ("The New England usage of blueberry for species of *Vaccinium* and huckleberry for the *Gaylussacias* is best," he wrote), which is confusing to a twenty-first-century reader only familiar with huckleberries as a signifier, mascot, and value-added product of the Northwest, where *Gaylussacia* is not a huckleberry, no way. The name game continues today: the 2019 Baker Creek Heirloom Seeds catalogue features a species of purple-black, berry-sized *Solanum* that originated in Africa. This berry is related to tomatoes and eggplants, not blueberries or heath, but Baker Creek calls them "garden huckleberry."

The genus and species names can be hard to follow, I know. Blame their profusion on a clash between plants that gleefully mix genetic material and Linnaean taxonomy's drive to categorize and name each offspring. Scientific names go through a community vetting process that can split one variety into two if scientists find that their DNA is different enough, or join two into one for the opposite reason. At the same time, common names multiply, giving one thing multiple names according to different cultures, even those who speak the same language. In English-speaking areas, for example, daffodils are also known as jonquils and narcissus, depending on where they bloom. Or common names are informally weeded from casual usage until we rely on one name for many different species, like huckleberry.

If you burrow further into huckleberry classifications, you'll find the two major species available for consumption today: *Vaccinium ovatum* (evergreen huckleberry, found west of the Cascade Mountains, dominant where I'm from) and *Vaccinium membranaceum* (mountain huckleberry, found east of the Cascades, dominant where I now live). These huckleberry fields are, some botanists say, part of a vast field of huckleberries that covered North America before geologic changes reduced them to berried islands clinging to mountainsides.

And what about *Menziesia ferruginea*, or "fool's huckleberry"? Not a berry at all, but a plant with leaves and flowers that look huckleberry-like and host a pink, berrylike fungus,

Exobasidium vaccinii, that grows on the undersides of the leaves. According to Betty Derig and Margaret Fuller's *Wild Berries of the West*, "Some tribes of the Northwest Coast ate the fungus berries, which are apparently not poisonous." (The rest of the plant is.) "The Tsimshian of coastal British Columbia ate the fungus even though they believe they were the snot of Henaaksiala, a mythical being who stole corpses."

In English, the word "huckleberry" is probably related to "whortleberry," meaning "little berry," a name first hung on the hedge berries of England, many different types of hedges and many different names, but with a common quality: little berries produced not as a crop but as a by-product of the plant's role as a boundary.

In the nineteenth and early twentieth centuries, a common use of "huckleberry" was as a way to express personal humility. "You're a persimmon over my huckleberry"—meaning, "You're bigger/more important than me"—hasn't aged as well as "You're a peach," but I could see plain old "huckleberry" making a comeback as a term of endearment. The littleness, the dearness, the lesser-than-ness, the wildness—that's what Twain wanted for Huck Finn's name. When Val Kilmer as Doc Holliday says, "I'm your huckleberry," in the movie *Tombstone*, he means both "I'm your guy" and "I'm gonna cause you trouble."

In Salish, the word for huckleberry is *sťxaɬq*. "*Sťx* is 'sweetness,'" LaRae says, "and *aɬq* is 'sweet smell.'" To have "huckleberry eyes" means your peepers are sweet, dark, and round.

///////////////////////////

Every Thursday from April to October, Mo unloads his Ford Aerostar in the parking lot of the Shop, the hub of Spokane's Perry District, and sets up his foraged-food booth next to my table of pastries. We greet each other by taking turns grabbing the legs of our polyester canopies and muscling them into place. We can put our canopies up alone, but it's easier with help.

Mo is more hard of hearing than I am—so much that I've stopped noticing how loud I have to yell to get his attention, since this is the only way we can talk. When I shout-ask how much his elderberries are, he startles like I just snuck up on him.

In May, Mo sells morels; in June, he sells porcinis. In July, there are raspberries in aqua-colored crates, and blue-purple sea holly dangling from the buckets he fills with water and ties to his canopy to keep the wind from tipping his whole operation over. In August, he sells perfect purple-black huckleberries, though lately they've been arriving earlier and earlier—just after Fourth of July this year, thanks to all the sun we've been getting and last year's forest fires. He is the market's most dependable source of berries year-round.

Sam does not like Mo. Has refused to buy berries from him since Mo snarled at him for questioning his huckleberry prices. "Ninety dollars a gallon? Really?" Sam had asked. "Best price you'll find," Mo said. "Don't you know it's been a bad year?"

The bad year was the summer when forest fires so completely circled Spokane that we couldn't leave our houses.

Like a blizzard, but of ash. One million acres across the state were consumed by fires so hot, in weather so dry, that they couldn't be extinguished until it snowed in October. Three firefighters died in a blaze near Twisp. A dark scrim of soot fanned from a window we forgot to shut and spread over everything the breeze could reach. Cans of ash from the 1980 Mount Saint Helens eruption that I'd set as decorations on the windowsill were smeared with ash from the new fires.

The year after those fires was one of the best huckleberry years in local memory. The berries, like the heat, came early and stayed. We bought gallons of them for $60, sometimes $50, feeling lucky to pay so little. Sam ate huckleberries on his granola every day—such an extravagance!—and made his huckleberry pie, just huckleberries and sugar and a little lemon wrapped in pastry, nothing to distract from the star of the summer, those deep, sweet, plentiful hucks.

When I ask LaRae to teach me how to say "huckleberry" in Salish, she says, "Close your eyes. I'm going to say the word and you're going to listen. Then say it back to me."

I am hard of hearing. In conversation, if I can see a face I can hear better. With my eyes closed, without facial expressions or the speech of eye contact, I am afraid I'll mishear.

But it's not like that. Instead, with my eyes closed, the buzz of the café fades, and I tune in to LaRae's voice, willing myself to feel comfortable in this posture of reception. Her voice is the only thing. She's saying *st̓xalq̓. St̓xalq̓. St̓xalq̓.* To catch the sounds, I imagine they are topography. This one has a steep incline, a sharp peak, then a shallow descent with a quick hitch up at the end. "Now you," LaRae says. I fumble through, and she says, "Good,

try again." This is one of the methods she and her teachers
are using to revive Salish—repeating words, playing games
with them, as you would with any child, until the words
lodge themselves next to English.

It is hard for me to get the sounds right. I'm embarrassed by this. I wonder if this tiny shame is a boundary I
have momentarily crossed, if I've moved from a place where
I am comfortable and fluent into a place where I am an out-
sider and strange, possibly suspect, definitely illiterate and
(in the old-fashioned sense) dumb. She says it again, *sťxałq*.
I try to master the air between *sť*, which sounds the way an
English speaker would expect, and *xa*, which sounds like
ash. Then the front-of-the-throat "k" sound of *łq*. "We are
children in the language," she says. "We are learning, and
we will get better. Mistakes are part of that, and nothing to
be ashamed of."

In the car, leaving our interview, I try to say the word
again and I can't. It's gone. I can't even recall how it begins.

But I can remember how to say "medicine." *Mrímstn*. A
little trip on the *r*, so you skip directly from *m* to *r* to the
"eem" sound, then a new syllable, familiar this time: "stn."
When trying to pronounce Salish, English speakers might
imagine an *i* in there—"stin"—but the sound is shorter,
acutely angled from the *t* to the *n*, not a full *i* (like "pin")
sound. I imagine this as a hollow my tongue steps over.
When I can't remember the Salish word for "huckleberry,"
I say the Salish word for "medicine."

When Sam first moved from Long Island to Spokane, he
saw a McDonald's sign that read "Huckleberry shakes are
back!!" and thought, Holy shit, I'm in another world.

Huckleberries are, to him, not just western but Western, the direction he was learning to call home. In Spokane, he could buy huckleberries in the freezer aisle of the fancy grocery store, fresh at the farmers' market, and swaddled in milk from a drive-thru that would also sell him a burger and fries.

Huckleberries evoke the West in packaged products that can be bought on summer vacations at Evel Knievel Days in Butte or ski resorts in Sun Valley, or eaten in pancakes at Frank's Diner in Spokane. Bear-shaped huckleberry honey jars, bear-logoed huckleberry wine, huckleberry syrup, huckleberry fudge, huckleberry taffy and jelly and jam— when you see these in a gas station, you know you're in the West. The hardest thing to find, if you aren't local, is actual fresh huckleberries.

In 'Asta Bowen's *The Huckleberry Book*, she uses poetry to explain their lure. "The huckleberry is wildness in your hand," Bowen writes. When you eat them, you take wildness into your body. "Imagine it storming your veins all day, coursing your heart like a western river, lining your bones with what gives the grizzly its grunt."

Bowen invokes what Sir James George Frazer would call sympathetic magic, a way of making meaning through metaphor (in the sense that metaphors aren't just units of language we learn about in poetry class, but also habits of everyday thought that move meaning from one thing to another). Bowen's magic is contagious magic, whereby objects, substances, and elements can transform the person who wears them, touches them, eats them, etc., by giving that person the characteristics of the element in question.

Like eating steak to feel manly, or wearing diamonds to feel forever loved. Like those of us who feel food is *mrímstn*.

This is not the kind of magic Frazer calls imitative magic, whereby a tree planted in the name of a dead relative whose roots are nicked by a lawn mower might make you feel like you hurt the person the tree represents, and rip your grief back open.

I am not suspicious of contagious magic. I am suspicious of how wildness can become a marketing ploy—a McDonald's marquee, a jar of tourist-trap jelly. A sales pitch that uses our craving for what's deep and real about the places we're from to sell an easy, sweet product.

But I also love that kind of bullshit. Maybe it's capitalism, but there is a wholeness—or a modern truth, at least—to having the authentic thing that's carefully gathered exist in parallel with a commodified version of itself. One preserves a true connection to our land. The other takes the fruit that signifies that connection and makes it accessible to everyone, regardless of whether people understand the history and effort that attends each berry.

Huckleberries are struggle and sweetness. To all of us who live in huckleberry country, whether we pick our huckleberries or buy them, they taste like home.

The first time I meet LaRae is at an event where she and Chris tell their life stories onstage. While LaRae speaks Salish, Chris translates. Then they switch—Chris speaks Salish and LaRae translates. I can tell which audience members speak Salish because they get all the jokes before I do, a ripple of laughter that makes me anticipate my own. This exchange of language is intimate and expansive at the same time. The closest thing I can compare it to is a bilingual Easter Mass in which my childhood community, split between English and Spanish speakers, came together to worship at an altar they did not usually share, carried there by a story of suffering and redemption told slowly, taking turns in everyone's tongue.

Salish is endangered because LaRae's grandmother's generation was "taken from their land by force of law and taught shame"—and English. "When I started learning my language," LaRae says, "I became complete. I think and I hope that's how it will be for our students and their families."

"Language is a way to respect the heritage of this place," says Chris, who identifies as white and, like his wife, is a polyglot—fluent in English, Spanish, and Salish, which he learned alongside LaRae. He asks us to imagine Salish being spoken on the street, in restaurants, at the bank and pharmacy and café, bringing our region's original culture back into earshot and into the mainstream present, restoring a wholeness to our city. He's not talking about the whitewashing of an imagined shared heritage. He's talking about having Spokane become visibly and audibly the place we are, the home that European settlement carved out of Native land, where the descendants of both peoples—plus transplants like Sam and me—make their homes.

Should domestication of the huckleberry for crop production one day succeed, it would breed away the quality that makes these berries so special: their interdependence with each other, their mountain environment, their stewards, their place—all the elements that give huckleberries their flavor and their importance.

Near the end of the evening, when the story is almost over, LaRae picks up her hand drum. She says in Salish, then English, "This song is called 'My Great-Grandparents Got Me Ready for This, My Grandparents Got Me Ready for This.'" She starts the steady thump of the beat, opens her mouth to the music, and sings.

Huckleberry Pie

Sam makes this pie whenever he can, usually in August, when huckleberries are fresh, and for Thanksgiving and Christmas in place of pumpkin pie, when he uses the last of our frozen berries. Baking with hucks we foraged ourselves would charge our pies with more homegrown significance, but the truth is, both Sam and I hate hiking. We usually buy or trade for our berries. To regain any contagious magic lost by this failure to pick our own, Sam makes a flaky American-style pie pastry with all butter or half butter/half lard, adding only the amount of sugar and spice needed to frame the huckleberries' wild sweetness.

Yield: 1 pie (8–10 servings)

- 1 recipe your favorite double-crust pie dough (or see page 317 for whole wheat pie pastry)
- 6 cups fresh or frozen huckleberries
- 1 cup granulated sugar
- Juice of ½ medium-sized lemon
- Big pinch of salt
- ¼ cup all-purpose flour
- 1 tablespoon instant tapioca
- 2 tablespoons unsalted butter, chilled, cut into small pieces
- Egg-white wash (1 egg white beaten with 1 teaspoon water)
- Demerara sugar, for sprinkling

Make the dough and refrigerate it for at least 1 hour, or up to overnight. Roll out the bottom crust, and place it in a 10-inch pie plate. Tuck the crust into the plate, trim the edges, and refrigerate. Roll out the top crust, fold it into quarters, and lay it on top of the bottom crust to cool in the refrigerator while you prepare the next steps of the recipe.

Preheat the oven to 425°F.

In a medium bowl, combine the huckleberries with the granulated sugar, lemon juice, and salt. Taste and adjust the flavors as needed. Gently stir in the flour, tapioca, and butter until evenly distributed.

Remove the crusts from the refrigerator and pour the huckleberry mixture into the bottom crust, and smooth it into a mound with your hands. Drape the top crust over it, trim the edges, and crimp or flute them. Cut generous steam vents, brush the crust with the egg-white wash, and sprinkle it with the demerara sugar.

Bake for 10 to 15 minutes, then check to see if the crust is blistered, blond, and no longer wet-looking. Lower the temperature to 350°F and bake for another 40 to 45 minutes. When the juices burble viscously at the edge (not merrily or liquidly), the pie is done. Cool on a wire rack until you can touch the bottom of the plate without burning your hand. Serve then, or once the pie has cooled completely (at which point it will be finished setting up), and eat within three days. To store, cover the pie with a dish towel and leave it on the kitchen counter.

Huckleberry Gastrique

Prepare this savory-sweet-tart deep-purple sauce when summer huckleberries are ripe, and freeze half of it for winter feasts. Spoon around discs of roasted squash or root vegetables, or drizzle atop meats. If you have access to elk or venison, try this on a slow-cooked sliced roast.

Pay close attention to the particular qualities of your huckleberries. Are they juicy? Sweet? Acidic? Under- or overripe? Varying rainfall and sun exposure will change the characteristics of berries from year to year; huckleberries picked in one place may vary from huckleberries picked elsewhere. Adjust the sugar and vinegar in this recipe to suit your particular berries.

Huckleberry seeds are almost imperceptible, so this recipe calls for blending all the ingredients together at the end. If you want to adapt this gastrique for another berry—say, Himalaya blackberry—instead of blending the sauce, strain it through a sieve to catch the seeds.

For those who revere huckleberries, it may be strange to do as I ask and blend them to a smooth sauce. To preserve the integrity of the berries, hold back some hucks to scatter over the meat before drizzling all with the gastrique. Be careful not to sweeten the sauce too much, or sweetness might overwhelm the huckleberry flavor. When selecting a red wine for this recipe, choose one that isn't too sweet, and start with the lower amount of sugar called for, adding more at the end if necessary.

Yield: about 1½ cups

    2 tablespoons unsalted butter
    1 shallot or ½ onion, minced
    2 cups plus 2 tablespoons huckleberries
    2 tablespoons sugar
    3 tablespoons red wine

3 tablespoons fruity vinegar, like raspberry or
    apple-cider
Big pinch of salt

In a saucepan over medium heat, melt the butter, then add the shallot or onion and sauté until it's translucent, about 5 minutes. Add the rest of the ingredients—except for 2 tablespoons of huckleberries—and bring them to a boil, stirring to dissolve the sugar. Reduce the heat, and maintain a low simmer for 10 to 15 minutes, until the sauce looks silky and a little thick and the huckleberries are very soft. Taste. Add more salt, sugar, or vinegar to your liking.

Transfer the sauce to a blender and purée.

Serve hot with a hot dish or cold with a cold dish. Scatter the reserved fresh huckleberries over the meat or vegetables you're serving, then spoon the sauce over the berries. Store leftover gastrique in the refrigerator and use within the week.

To freeze the gastrique, ladle into a container of your choice, leaving plenty of headspace.

I: Italian Plum

Prunus cocomilia
Rosaceae (rose) family
Also known as Empress plum,
Italian prune plum, prune tree

I doubt that my sort of nostalgia is what Jane Grigson meant when she wrote in her *Fruit Book*, "We all probably have some kind of romantic view of plum orchards," though she does include "the slow murmur of wasps" in her vision of mossy, crooked branches heavy with red and gold drops.

I think *plum* and remember the summer before I turned twenty-nine, about a month before the Italian plums in our backyard would ripen. They were hard clusters of green fruit then, each one a promise of the preserving frenzy to come.

I'd decided that if I didn't leave W immediately and

without warning I'd never leave him at all. Later, I'd explain my decision with gardening metaphors—after years of collecting seeds and tools and dreaming of the harvest I planned to raise like a miracle from the earth, I couldn't face the need I'd created. An actual gardener might feel the dread of her chores gradually and abandon only parts of her ambition, still raise the easy stuff, the bush beans or rhubarb or radishes, pick the plums that would arrive regardless of her care. But W was a person. I couldn't imagine tending him in half-measures.

I packed a suitcase, hid it in the trunk of my car, and waited for him to return from a recording session. Then I told him we were through. He said he was going to get famous and I wasn't going to be there to see it. Later, when I wanted a public shorthand for why I left, I'd retell that part. My private shorthand was another gardening metaphor.

Our plums.

They were ripe when I returned to get my things. Each time I filled a cardboard box, I took a break to eat one. Firm, sweet, not so juicy I had to be careful of where I enjoyed them. I didn't even need to wipe my hands when I was through, which was good, because I'd already packed the dish towels. I have no doubt the rest of that year's fruit fell into the grass. A loss, but not a tragedy. With or without us, the plums would return the next year.

Italian plums are so much sweeter than this memory suggests. On this, I'm sure Jane G. and I agree. On this, too: plum trees bear more than fruit. To have one in the backyard—and in that Seattle neighborhood at that time it seemed like we all had Italian plum trees in our backyards—said that,

though I had turned our house into just another rental, it
had not always been that sort of house. At some point, people
had lived here who had planned on staying. I knew because
they'd planted plums.

There were cherry and pear and apple trees in that
neighborhood, too. They all seemed like ways to read the
urban landscape and imagine the people who came before,
the dramas they'd lived before our dramas, the bounties
they'd planned until plans changed. But, unlike the other
fruit trees we inherited with our yearlong leases, Italian
plum trees bore delicious fruit whether we took care of
them or not. Their low maintenance mattered, because W
and I were renters and our friends were renters—that Seattle neighborhood even then being too expensive for us
to dream of permanence, but not yet so expensive that our
paychecks couldn't cover a group house and a recycling bin
jeweled with glass fifths. We'd almost figured out how to
take care of ourselves, but no way did we know how to take
care of trees.

I did know how to take care of W. He required rides to
and from work, and to borrow my car for band practice.
He required me to listen to his songs and tell him they
were good. He required whiskey, then vodka, then aqua-
vit. He required space, no cuddling, being in too much
stomach pain to stand any physical affection. He required
me to be there. He did love having me around. Never happier than when he was hung over with no reason to leave
his room and had me beside him, reading. He required
elaborate meals I loved to make, pudding cakes and cornbread like his mother baked. Then he required everything
wheat-free, soy-free, corn-free. No onions, ever. No milk if I

could help it. One day near the end, he decided he required a completely gluten-and-dairy-free diet. I remember that he seemed nervous about telling me, and that I received the news with a long face, feeling like my world had gone gray. He never ate the fruit from our backyard. The plums were not his thing, except as a backdrop for watching me tend my rows of tomatoes and peas, a sight he said was the most beautiful thing and made him feel like we were a family.

Then and now, the plum trees of the neighborhood were plain, with plain foliage that made a plain yard look like someone's interrupted dream of a garden, usually planted against a back fence, which guaranteed a little fruit for the neighbors. The moment we noticed their papery blossoms, they blew off in the rain. After that, we'd forget about the plums right up until their fruit startled us in late summer by turning a deep, demanding purple. Picking plums into a giant canning pot wasn't smart—fruit at the bottom bruised from the weight of the plums above them—but I did it anyway, my yearlong neglect of these trees mutating into days of urgent greed. Even when the plums were bruised, even when they had worms, I could use them. The bad sections were easy to cut away.

Do you already know what Italian plums look like? They have purple skins with blue fuzz, and firm yellow flesh. Also called Italian prune plums, they're a freestone variety easily parted from their pits, one trait of several that make them ideal for baking and preserving. Comparatively high in sugar and acid (in plums it's usually one or the other), they keep their shape and deepen their nuance when cooked, unlike their more straightforwardly sweet cousins, such as Santa Rosa or Mirabelle or Elephant Heart, which dissolve

into petulant juice. When raw, they are not as affable as plu-
ots or as complex as greengages. When dried, they shrivel
into handsome black prunes. In early September, they seem
to ripen all at once, and there are always too many of them.

An Italian plum tree is fruitful for twenty to forty years,
which means the plums I remember were probably planted
within my lifetime. They could have been donations from
the city of Seattle, which has run tree-giving programs like
the Department of Neighborhoods' Tree Fund for years, but
it's more likely they were purchased from a local nursery.
"Our goals are canopy-oriented, and fruit trees typically
don't contribute as much to the canopy," says Lou Stubecki,
the manager of Trees for Seattle, the current Department
of Neighborhoods program that gives free trees to city
residents. I had imagined, until I talked to Lou, that my
neighborhood's profusion of fruit-bearing trees must be
the consequence of a civic program, that the plums and
cherries and pears were part of a larger plan, the leavings of
a fruit-positive culture that owned these houses before we
renters arrived with our laptops and guitars.

There was no such plan. The city did for an indeterminate length of time give a limited number of fruit trees
away, but as of 2019 they do so no longer. "We're not down
on fruit production," Stubecki says. "We just don't want to
be involved in it." Fruit trees invite pest problems that residents might respond to by irresponsibly using pesticides,
he says. Plus, when not properly cared for, they make a
gigantic mess.

Homeowners can buy trees and plant what they want
on their own land. But when a house passes into posses-
sion of someone who doesn't want to or can't take care

of the property's fruit trees, I can see why municipalities don't encourage planting more of them—uncared-for trees drop their harvest on sidewalks and streets, over roofs and lawns, weaving a rotten carpet of pulp and pests. In Seattle, residents can call volunteers at City Fruit, a local nonprofit, to pick their neglected crops, and City Fruit will give what they pick to food banks, preventing the trees from becoming a nuisance and connecting hungry people with fresh homegrown fruit. This is an ingenious patch to the missing link between those who plant trees and those who inherit them.

But why do we let our fruit become a nuisance in the first place? Because so many of us are accustomed to buying fruit pest-free, in easily handled quantities, from twenty-four-hour supermarkets? Or do we decide this is not the thing we need, all this fruit and the care of the fruit and the eating of the fruit we cared for, the entrapment of all this nurturing? In the land of plenty where I lived with W, with two grocery stores within walking distance, fruit trees were superfluous but joyful luxuries. There were more of them than there were people who knew how to care for them. University extension programs and organizations like City Fruit offer cheap or free classes to bridge this gap of horticultural knowledge, but an important need remains unmet: fruit trees produce food, they usually require special care, and what makes all the fuss worthwhile is eating the food, which requires knowledge of how to prepare and preserve it. If you lack the ability to jam, pickle, dry, or ferment, an Italian plum crop is a novelty in the first week, an anxiety the second, a mess in the third, and by the fourth, a burden.

Fortunately, Italian plums are patient teachers when it

comes to preserving. In jam they are slow to scorch, easy to set, with a pleasant zing of acid to balance their modest natural sugars. They accommodate vanilla, cardamom, clove, thyme, almost any flavor you crave. Their pits are easy to remove and make an almond-scented noyau like the "almond" extract on page 40. There's no need to peel these plums, for their skins are thin and flavorful and turn plum preserves an appetizing fuchsia, which means that prepping a plum recipe is a matter of a quick chop. It's hard to fail at plum jam, but should one fail, there are always more plums to succeed with.

Jamming the harvest will, I promise, become a chore. Once it does, you can finish the rest of the plums by drowning them in liquor to make slivovitz or drying them under a hot sun to make prunes. Or—what I would suggest as the grateful recipient of my neighbors' excess—you can give the plums away.

When I said W required elaborate meals, I mean I enjoyed making those meals and assumed he enjoyed eating them. When I said he required liquor, I mean drinking with him was fun. When W said, after I left, that he'd never asked me to help him, he was technically right. I did what I thought a good girlfriend would do, conflating care and need with love and desire. I did not ever literally think this. I tended him, at that time, without thinking, as if there were no other way to love.

As I write this, I live on the other side of the state in a house with a small backyard and a lawn that ejects neon rocks whenever we dig into it. I've been fussy about planting fruit trees, worried we won't pick the right place or keep them alive. I should have done as our new neighbors did, as European pioneers to this area used to do—plant trees immediately, to signal we're staying. Instead, I watch the market. When Italian plums appear next to the season's first apples, I call my neighbors and friends. Someone is always overwhelmed with their feast of plums. And I am always ready to help.

Italian Plum Jam

In this jam, Italian plums are cooked whole into a dark-magenta preserve. It can be used in either a sweet or a savory pairing, on yogurt or brie, dolloped between cake layers, spooned onto bread and crackers with cured meats and aged cheeses, or eaten straight from the jar. It is possible to use up a whole harvest with this recipe, though you'll probably get sick of standing over a hot stove before your plum supply runs out. If that happens, remove the pits from the remaining plums, place the fruit on metal baking sheets, and leave them in the hot sun until they prune. The metallic shine of the sheets will scare birds away; flies will check the plums out, but in my hot-dry summer climate, they do not affect the final product. The amount of time the plums take to dry will depend on the weather, so check on them once a day. I like to take them inside once they look wrinkled and black but before they desiccate, so their sugars are concentrated but their flesh is still soft. Store these prunes in the refrigerator—they'll mold if forgotten in the pantry at room temperature.

This plum jam is the best of all the plum jams I've ever made, brought to my attention in *Cather's Kitchens*, Roger L. and Linda K. Welsch's study of food in Willa Cather's fiction and life. I've modified a recipe they found "between the pages of the Cathers' *White House Cook Book* . . . torn from page 9 of the September, 1909, issue of *Woman's World*," interpreting cooking instructions to suit a modern stovetop (the original recipe called for putting "the plums over the fire"). I also added lime juice because it's delicious, but you can substitute lemon.

This recipe can be doubled, but even after years of practice my best results are with 1-kilogram batches.

A note on obtaining plum kernels: you really need a mortar and pestle to crack plum pits open. A hammer will do, but a mortar and pestle is the best tool to extract the kernels intact. This takes precision pounding—not so soft that it has no effect, and not so hard that it pulverizes the kernel in the process of freeing it from the shell. Be prepared for flying shells. Keep extra pits—you'll smash some kernels in the process of extracting them. Keep the smashed kernels for persipan (following recipe) and keep cracking until you have the right amount of kernels. I have aesthetic reasons for this. If you don't care about smashed kernels, just throw them in. You may want to wear goggles—or just shut your eyes at the moment of contact. A broom at the ready is a useful tool.

It may also be helpful to count the number of plums in your batch before cooking them, so you know how many pits you'll need to retrieve from the jam.

Yield: about 32 ounces

1 kilogram (2.2 pounds) Italian plums, washed but
   not pitted
¼ cup water
750 grams (almost 3½ cups) sugar
Juice of 1 lime

Place a small plate in the freezer; you'll use this later to test the set of the jam. Then prepare a deep canning pot with enough boiling water to cover five 8-ounce jars. I add a little white vinegar to mitigate the powdery white residue my hard water leaves on my jars after I boil them. Sterilize the jars by keeping them immersed in the boiling water for 10 minutes, then set them on a clean towel to cool and dry. Boil the lids for 10 minutes, then place on a clean, lint-free towel, seal-side up, to dry. Rinse the bands and set them aside. Keep heating the pot of water on low until the end of this recipe, when you'll use this water bath to process the filled jars.

In a preserving pan, cook the plums over medium heat, adding the ¼ cup of water at the beginning, until the water sizzles and the plums have warmed and softened. Then press each plum with a muddler or a wooden spoon to help it release its juice. Let the plums boil gently for 45 minutes, stirring and smashing them every now and then to help loosen the pits. Reduce the heat as needed to keep the plums at a gentle boil. By the end of the 45 minutes, the plums should have collapsed entirely into a bright juice, and the pits should be loose in the juice or plum flesh.

Remove from the heat. Fish out all the pits, and set them aside. Add the sugar and lime juice. If you wish, you can let the mixture macerate in the refrigerator overnight. (It makes no difference to the final results, but sometimes it's nice to take a jam break.)

Wash the plum juice from a quarter of the pits (seven or eight, probably), and crack them open with a mortar and pestle, retrieving the kernel from the center of each pit. Peel off the brown skin of each kernel (this can usually be done with a firm pinch at one end of the kernel). Set aside.

Return the preserving pan to medium-high heat, and boil the plum-sugar mixture for 8 to 10 minutes, stirring only occasionally, to prevent scorching. I start the timer once the jam is at a full boil. Skim and discard any fruit scum that appears. At minute 5, stir in the plum kernels. The jam is done when it is dark and glossy and will cling to a spoon after a minute at room temperature, or wrinkle when you push your finger into it after a minute on a plate in the freezer. Skim the last swirls of scum and any pits that floated to the surface and pour the jam into jars, leaving ½ inch of headspace. Wipe the rims clean, place the lids, and screw the bands on, fingertip-tight (tight, but not so tight you need strong hands to reopen them). Bring the water bath back to a boil and process the jars, keeping them immersed for 10 minutes. Remove the jars from the water, and set them on the counter to cool. Store in a dark cupboard at room temperature, and refrigerate after opening.

Persipan

For a marzipan-adjacent paste that's almondy, rich, and bitter-sweet—or to ascertain if you really are the sort of person who adores detail-oriented, repetitive tasks—try making persipan. Usually it's made with apricot kernels, but any stone-fruit kernel can substitute. All stone-fruit kernels contain cyanogenic compounds that should be treated with caution, but not fear. Roasting the kernels will "de-bitter" these "bitter" almonds (remember, they are not true almonds, and the bitterness refers to removing cyanide, not the bittersweet taste of each "almond"). Most recipes recommend not browning the kernels, but browning them can add a delicious toasty flavor. It's your choice.

Persipan will be coarser than marzipan and not as amenable to forming fancy shapes. Use it instead to line the bottom of a plum galette or a peach pie, or eat the persipan straight from the jar by the knifepoint, savoring it as you would a piece of dark chocolate. I find it useful to save stone-fruit pits through the summer, until the end of the season, then invite a friend over to help me smash all the pits at once.

Yield: varies

Plum, peach, or apricot pits, as many as you can find
Sugar
Pinch of salt
Water

Remove the kernels from their pits by smashing them with a mortar and pestle or with a hammer. Discard the pit shells. Blanch the kernels in boiling water for 15 seconds, then peel them by giving each one a hard pinch at the edge and slipping the peel off. At this stage, their almond scent will be mixed with a starchy scent, like overripe peas.

Preheat the oven to 355°F.

Arrange the kernels in a single layer on a baking sheet, and roast them in the oven for 10 minutes (the cyanide breaks down at 354.2°F). The kernels shouldn't toast, but if they do, that flavor will be part of the persipan, which isn't a bad thing! Your kitchen will begin to smell like almond cookies.

Cool the kernels, then measure them. Divide their volume by three and multiply that amount by two to find the appropriate measure of sugar (60 percent kernels, 40 percent sugar). If you have 3 cups of kernels, for example, you'll add 2 cups of sugar. If you have 1 cup of kernels, you'll use ⅔ cup of sugar.

In a food processor, combine the kernels and sugar with a pinch of salt, and blend until the kernels are finely ground. Add water, a teaspoon at a time, until the persipan comes together into a paste, blending in the food processor the entire time. When the persipan starts to cling to the blade in a manic ball, it's done. Stop adding water and stop blending.

Store in a covered container in the fridge. As far as I can tell, persipan will keep indefinitely.

J: Juniper Berry

Juniperus communis
Cupressaceae (cypress) family
Also known as common juniper, savin

This is none of my business and not my story, except I was there, fresh off the flight to Denver, sitting with J while the doctor narrated the procedure and her machine whirred, suctiony, sounding just as I'd imagined. J's hand was cool and thin and foreign, the physical affection we might give each other so assumed that we usually save it for only our most intense moments. We've always been almost exactly the same size—her shoes fit me, but too tightly to walk very far, even when we were twelve—and when we were eighteen, neither of us could reach the nine-note chords the next stage of our piano training required. I'd interpreted this as a sign that my body wasn't built for that kind of music and quit. She kept playing.

While J grimaced and closed her eyes and got through it, the doctor spread a thin layer of chat over the other sounds in the room—the weather, but not our weekend plans. I brought up pie. Everyone has something conflict-free to say about pie. I know because I've written a pie cookbook and I use this sort of conversational fat often. Oh God, I thought. Here I go again, talking about myself when J should be the center of attention.

Still, she *was* the center. The woman doctors and sister-friends circled around this sculptor and piano player and caregiver and cook, this single woman not ready to be a mother. We wrapped her in blankets of butter talk and suction sound. I squeezed her hand, but by then the cramps the doctor promised had started, and J didn't squeeze back.

Then it was over. Faster than I'd thought. She was pale and didn't want to talk. I brought her car around and took her home.

In eighteenth-century England, gin earned its nickname "mother's ruin" not only for its inebriating effects, its addictive properties, and its reputation as a woman's drink, but because juniper, its primary botanical ingredient, contains terpene-rich essential oils that thin blood, regulate menses, and in higher quantities cause miscarriage. Juniper has its own nicknames and aliases: savin, bastard killer, emmenagogue, menses regulator, menses bringer, and the clearest of them all, its meaning apparent in English at the Latin root—abortifacient.

My grandfather said that the only picture he had of his mother was in her coffin, that she'd died of measles when he was four and his stepmother never loved him right. The last time I saw him, he, too, had died, and we hadn't found the picture of Great-Grandmother Edith. When my parents left the viewing parlor, I took a photo of him, my iPhone lens still set to "square," the perfect size for Instagram. He would have looked like he was sleeping if the ruffles of the mortuary bed hadn't given him away. "The sooner you get here the better," my father had suggested, not because he was telling me his dad was dead and he needed comfort, but because my grandfather wasn't embalmed. "Measles! Where'd you hear that?" my elder uncle had said when he thought I was old enough to hear the rumor. "His mother didn't die of measles. She died of an abortion."

Literature on juniper as a culinary spice is easy to find. The berries are described as resinous and piney, with a sharp aroma that makes them a suitable substitute for black pepper and a flavor that complements cured meats, pickled fish, sauerkraut, game, and other savory dishes popular in Western and Eastern Europe—though juniper flavor will be most familiar to British and American drinkers as gin.

Juniper is also easy to find in old herbals, but its appearance is almost always accompanied by a reticence about the herb's relationship to birth control. *Gerard's Herbal*, published in 1597 and popular throughout seventeenth-century England, is more specific than most about juniper's uses for "women's troubles," but Gerard does not record ways to prepare or administer juniper as medicine. He writes that the leaves, when "boyled in Wine and drunke," can "bring downe the menses with force, draw away after-birth, expel the dead childe, and kill the quicke." He does not explain when or how to take the drug, nor does he describe useful doses. Gerard writes "kill the quicke"—the moment when the fetus, now ensouled by God, moves within its mother— without qualification, unclear whether this note is a warning or a promise.

In 1653, the rebel English herbalist Nicholas Culpeper included juniper in his *Complete Herbal*, but he codes his abortifacient references, writing only that intake of juniper berries "helps the fits of the womb" and gives "safe and speedy delivery to women in labour." Culpeper wasn't shy about aggravating the medical associations of his time by making herbal home remedies accessible to laypeople, but he *was* shy about providing usable information about birth control. He might have been afraid of the law, which at that time punished wisewomen, doctors, husbands, and lovers for administering abortifacients if they knew the woman was pregnant, but usually did not prosecute the pregnant woman. Culpeper may have thought that abortion was immoral. Or that knowledge of herbal abortifacients shouldn't be common. The records left to us aren't clear.

This coded language is common in European herbals even into the twentieth century, as I find when I search Maud Grieve's *A Modern Herbal*, published in 1931 in England, the year my great-grandmother might not have died of measles in Omaha. Grieve is tight-lipped about juniper's association with women's health, mentioning that it is an emmenagogue—a term for something that "brings menses"—only in the last sentence, and only as an aside about a related plant, *Juniperus virginiana*, or red cedar, or pencil cedar, grown in the United States to make pencils and moth-resistant chests and squeezed of its oil for perfume and insecticide, oil that also happens to be diaphoretic (causing sweating) and an emmenagogue (something that increases menstrual flow). Emmenagogue is an afterthought, a little fact thrown in at the end. Did she do this on purpose, to be thorough but unemphatic? Does this subtlety imply that Grieve feared being caught giving abortion advice, or that she judged the abortion seeker and chose to deny her advice? Or does this brief reference mean what it appears, on the surface, to mean—that serving as an emmenagogue is a minor use of juniper?

Emmenagogues help regulate menses, which could simply mean getting a late period on track. Not an abortifacient, but another tool to maintain normal reproductive health. Emmenagogues could also be insurance against missing a period because of possible pregnancy—so maybe an abortifacient, depending on whether the woman was in fact pregnant, which she might not even know. They could also be used on purpose to cause miscarriage—in that case, definitely an abortifacient. It's all a matter of timing: an emmenagogue that would terminate early pregnancy might also be used in labor to aid delivery of a healthy child by accelerating uterine contractions.

Today, juniper remains scarce in modern herbals and popular botanical texts. It doesn't have its own entry in *The Master Book of Herbalism* (1984) by Paul Beyerl, or appear in *Nature's Pharmacy* (1988) by Christine Stockwell, nor is it foraged in *Stalking the Wild Asparagus* (1962) by Euell Gibbons.

In the *Encyclopedia of Folk Medicine* (2004), Gabrielle Hatfield gets direct with this passage, from her entry on abortion, not juniper: "Apart from purely mechanical means of destroying the fetus, in poor country areas of Britain, girls 'in trouble' would resort to strong laxatives or jumping off walls or stairs in an attempt to abort. If this failed, often gin was the next resort. Interestingly, gin is flavored with juniper (*Juniperus spp.*), one of the few plants that we know was almost universally used for abortion." Hatfield, like the sources she summarizes, does not provide details of how these women would have used gin or juniper, in what quantities and concentrations the tincture or tisane or decoction would have been prepared, whether women simply climbed into scalding-hot baths with bottles of home brew, or whether by "universally" she means worldwide or just throughout Europe. Hatfield hypothesizes that abortifacient recipes may have been obscured or left out because their authors feared persecution, or the recipes could have been lost. I wonder if they exist outside the text or over my head, like old cookbooks that assumed you knew how to cook and glossed over what today would take *Cook's Illustrated* three pages to explain.

I spend hours at the University of Washington's Miller Library, a horticultural archive, looking for overt connections of juniper to abortion within horticultural texts, and find instead mostly code words. "Savin" and "emmenagogue" are useful for index searches to find information scattered here and there among other plants and indications. The *Encyclopedia of Folk Medicine* notes that the subject has been taboo in Europe and North America for so long it is "difficult to know what the 'folk' actually did," though it's thought that midwives did maintain herbal abortion knowledge and have passed it down. Hatfield notes that the UCLA folk medicine archives, a treasure trove otherwise, is little help. Daniel E. Moerman's ethnobotanical records of Native American abortifacients, on the other hand, is so extensive that it suggests to Hatfield that abortion wasn't taboo to many tribes.

Moerman's *Native American Ethnobotany*, an encyclopedic collection of medicinal and culinary plant-lore, lists thirty-five tribes who include juniper in their medical traditions, a fairly long catalogue for the book. Among them is a note about a "compound infusion of [juniper] bark taken for women's diseases" by the Delaware of Ontario. This is the closest the juniper section comes to describing reproductive medicine—if we believe "women's diseases" is code for reproductive health. Moerman doesn't provide the context that would lead a layperson to conclude that abortion wasn't taboo to Native Americans, but my inability to draw conclusions doesn't mean Hatfield is incorrect. It's plausible that colonization would include wiping abortifacients—and whether or not they were taboo—from memory. Assimilation, enforced not only by language and land and blood but by a fresh infection of church-endorsed shame for having reproductive agency.

Is there any place and time when abortion has been a choice without threat of shame—just one possibility in the cycle of reproduction?

Maybe, John M. Riddle writes in *Eve's Herbs: A History of Contraception and Abortion in the West* (1997). He explains that the "quicke" I encountered in *Gerard's Herbal*, the moment when a woman could feel her child move within her womb, was at that time considered the sign that she was irrevocably pregnant. The weeks prior to this quickening were an indeterminate period when a woman couldn't be certain she was pregnant, which allowed her to keep her possible condition private. During this interval, she could take a potion to bring down her courses without having committed the sin of abortion, since there was no certain pregnancy—no state- or church-recognized life. After the quick, terminating a pregnancy was forbidden.

The quick remained pregnancy's defining moment in common law until nineteenth-century scientific advances revealed that the development of a fetus was not a set of definitive stages, but a gradual process. There was no clear line between not-life and life, pre-ensoulment and post-ensoulment. With this new knowledge, England in 1837 outlawed abortion before *and* after quickening, and in 1869 the Catholic Church abandoned the concept of ensoulment at quickening. This, briefly stated, is how we arrive at the idea contained in the rallying cry so well known to modern women, that abortion is wrong because "a fetus is alive from the moment of conception."

What a joke! The scientific advances that developed safe pharmaceutical birth control also closed a woman's window of indeterminacy, causing societal shifts that narrowed her control over her own body. Today, some drugstore pregnancy tests claim they can detect pregnancy five days *before* a woman even knows her period is late. There is now almost never a time when women can't be sure, one way or the other, what sort of choice she's making.

What I can't find at the university library is on proud display at my local anarchist bookstore. Their rack of zines includes *Crap Hound*, volumes 7 to 12, and *Backwoods Surgery & Medicine* (reprinted in the back by their in-house press), plus an offset-printed, full-color illustrated zine called *Reclaiming Our Ancient Wisdom: Herbal Abortion Procedure and Practice for Midwives and Herbalists* by Catherine Marie Jeunet, with an introduction by Esther Eberhardt. First printed by Eberhardt Press in 2007 in an edition of five hundred, the copy I hold is part of the 2016 printing, its fourth. It costs $10, no trade. I buy the last one in the store. As of this writing, you can order more online, where you will be warned, "This is not a DIY manual. If you need to terminate a pregnancy, we advise getting a medical abortion, which is safer and more reliable . . . However, we also believe this knowledge should be preserved."

The zine names herbs I know already—pennyroyal and rue and blue cohosh and tansy and Queen Anne's lace (but not juniper!)—and provides guiding doses. After unprotected sex, guard against pregnancy by chewing "1 tsp. of seeds twice a day for seven days"—a more specific instruction than anything I've encountered so far, but still not quite a recipe. If a decoction or infusion or tincture is called for, I'm lost. I don't know how to tell the strength of the medicine I'd be making, and I'd be too afraid of hurting myself or a friend to try. Perhaps I am not the right sort of woman for this knowledge. Not an herbalist. Not a midwife. Not a witch.

In the introduction, Eberhardt writes, "We have been forced to sift through dogma and moralism to find the information we seek"—as would be expected, given the intense political conflict around abortion rights, the shaming rhetoric, and the violence faced by abortion providers.

She goes on: "If an herbal abortion does not work for you, or is not appropriate to your situation, it does not mean you are less in touch with your body or your spirituality. An herbal abortion is no less 'right' or 'wrong' than a medical abortion."

Wait. What?

Read it again: "An herbal abortion is no less 'right' or 'wrong' than a medical abortion."

Is it possible that, even after she has sidestepped the shame of having an abortion, the way a woman chooses to terminate a pregnancy can be put on a scale of natural to artificial, good to bad? Can medical abortion possibly be less self-loving, less moral, than a "natural" pennyroyal tea taken at home without medical supervision?

Is there no decision we won't scour for purity? No place shame can't patrol?

Plants cannot tell you to wait seventy-two hours to be really sure, or require parental permission, or force you to face your embryo on an ultrasound screen, or force you to accept or refuse an offer to listen to its heartbeat, or force you to confirm that you have read a booklet that contains the phrase "The life of each human being begins at conception," before you are allowed by your state (in the case of these examples, Missouri) to follow through on your choice to terminate your pregnancy. Abortions may be banned from insurance coverage except in cases of incest, rape, or the endangered life of the mother, but plants often come cheap. Plants will not damn you to hell—though not knowing how to use them could maim or kill, and they can make a woman very uncomfortable even when used properly (as can medical abortions—"medical" becoming an adjective that pops up in herbal-abortion zines to distinguish between the procedures and pills directed by a clinic and the teas and tinctures of antiquity). Herb-induced abortions do not judge, the reason often invoked by the modern women who've used them. Still, their pursuit of an herbal abortion should not be read as a sign of radical independence without also being read as a sign of desperation.

For me to report that women "often" invoke fear of judgment as a reason for seeking an herbal abortion is a stretch. There are no reputable studies of the efficacy of herbal abortions, and no studies of how women feel before or after them. So strike "often." Strike any certainty about herbal abortions at all.

A few things I do know for sure: the juniper berry usually used in gin is *Juniperus communis*. Pregnant women are warned to avoid eating them in, say, gravlax or pork or beef dishes that use them for flavor. That pregnant women should avoid juniper-flavored gin (and hard alcohol of any flavor) is a matter of public health and well advertised, though I could not find advice on whether the juniper is more dangerous than the alcohol. Other species of juniper berries are probably at least somewhat toxic, but their toxicity hasn't been studied at length.

I know juniper berries take two to three years to ripen, so a bush never offers a pure field of ripeness. Pick the blue berries and leave the green for next year. When they're dried, their blue will turn a warm black.

They aren't berries, actually, but edible cones. Juniper bushes are dioecious, which means each tree is either male or female. Only the female cones are edible. When I put a pint of blue, probably not edible cones picked from a random hedge into an open Mason jar to sit on my desk as an altar offering to this chapter, they spring open a few days later, the dry heat of my house a signal to let loose. The oblong seeds rattle to the bottom of the jar, leaving their burst shells to curl above them like claws. For a week, the room smells like evergreen hedges before I throw the jar out, stuffed-nosed and headachy.

A male juniper tree can fertilize a female tree one hundred miles away. When these plants are enjoying their seasonal sex lives, I'm miserable, fog-brained, and nauseated. "Are you sure you aren't pregnant?" my mother asks when I call her during an allergy attack. She discovered her pregnancy with me in the process of answering an X-ray technician's routine questions, so finding a pregnancy while trying to treat allergies wouldn't be a strange turn for our line. But I'm not pregnant. Just allergic.

On a bad juniper day, I might irrigate my sinuses three or four times, relieved for a few hours before I need salt water and gravity to scour me out again. Since reading a story about the death of a Florida man by an amoebic brain infection he caught from his unwashed neti pot, I clean mine fearfully and only use filtered water, imagining each time I flush pollen from my nose the ways I could become a statistic: the woman killed by brain worms caught from DIY allergy treatment, the woman who hemorrhaged to death from pennyroyal. "NEVER TAKE ESSENTIAL OILS INTERNALLY," warns *Reclaiming Our Ancient Wisdom*, too late for the unnamed dead women it's using to scare us into caution. Even culinary advice hews this way; juniper's entry in Ernest Small's *North American Cornucopia*, a reference of indigenous food plants, cautions, "Only professionals should use juniper oil for culinary purposes—it is too concentrated and potentially poisonous." To be the woman who dies from the audacity of self-care—I imagine it and blow my nose and put the fear away.

I was raised—not on purpose, but as a by-product of my elders' encouragement to take charge of my own life—to believe that motherhood was the surest way to sabotage the achievement of anything else. "I wanted to be a nun, because nuns *did things*," my great-aunt Sister Maria tells me, and I think, as I have thought before, that if I'd been born in an earlier time this is the reason I would have been a nun.

Instead, I am a writer, have spent my adult years trying to be a writer, with all the attendant insecurities and uncertainties and failure, *plus* the terror that having a child would force me to give up writing, *while also* being pretty sure I wouldn't terminate a pregnancy. I've made choices of self-censorship, of abstaining from casual relationships, of bulletproof birth control, sex only with the feeling that I was nurturing an emotional exchange (though that, of course, was no guarantee of security). The riot-grrrl feminism that shaped my ideas about female power encouraged me to be loud, self-possessed, and confidently sexual. Unequivocally pro-choice. Freedom with my body was, I understood, a way to attain freedom from the patriarchy.

In the right circumstances, sex could be freeing, sure.

But it never felt free. It remained instead full of consequences, abortion an option for others but not for me.
Within my community, it was fine to be abstinent. That,
too, was a form of self-possession. Still—that choice, so
influenced by fear, felt more like repression. The church
I'd left and the ambitions I'd nurtured had conspired to
ruin sex.

Now, when friends call needing an ear because they're
pregnant and they're not sure they want to be, my response has been, "Anything you choose is the right choice."
One way, you have your life as you've chosen it, with time
to choose motherhood later—the choice that feels most
familiar to me. The other way, you have a baby, which I
can't imagine regretting. Or allowing myself to regret.
"Are you really this neutral?" J asked me after she recovered from her "termination," the word she prefers to use,
bracing to be judged even by the friend who held her hand
through it.

Juniper doesn't show up in old English texts as much as
savin does, a nickname for common juniper that also refers to savin juniper (*J. sabina*). It appears in a variant of
"The Queen's Maries," also known as "The Fower Marys,"
the sixteenth-century (or possibly eighteenth-century) folk
ballad written down by Francis James Child and attributed,
as so many of those ballads were, to Anonymous. "I would
venture to guess that Anon, who wrote so many poems
without signing them, was often a woman," writes Virginia
Woolf in *A Room of One's Own*. All versions of the ballad
share a narrator, often a Mary Hamilton, who sings,

> *Last nicht there was four Maries,*
> *The nicht there'l be but three;*
> *There was Marie Seton, and Marie Beton,*
> *And Marie Carmichael, and me.*

The narrator is pregnant by Queen Mary's husband and tries to abort her child. Having failed, she gives birth to the baby, then kills it. By the opening of the song, she's been caught for her crime and sentenced to death. In another version, our narrator turns first to the "savin tree," but the juniper doesn't work, as so many herbal abortions don't. Instead, Mary Hamilton throws her baby into the sea. Many versions imply greediness for the king's soft bed and his fragrant spices, a Mary who craves comfort more than sex—or maybe spices and softness are a coded way to refer to lust. I looked for coercion by the king, assuming Mary could have been pressured into sex, but if that's the subtext, I can't find it in the brogue.

There is no Mary Hamilton or Queen Mary that fits this story, historically speaking, but Mary Hamilton is real enough. Hers is the voice of women who have dreamed above their station, exercised their sexuality outside marriage, and are preparing to be punished.

Virginia Woolf speaks in Mary Beton's imagined voice throughout *A Room of One's Own*—a fact I'd forgotten, really only remembering the most referenced idea of her manifesto (to write lasting works of literature, a woman needs money and privacy and freedom from childbearing). "Call me Mary Beton, Mary Seton, Mary Carmichael or by any name you please—it is not a matter of any importance," Woolf writes, never mentioning the fourth Mary, the doomed Mary. We're supposed to uncouple Woolf from the "I" and let the "I" be fluid, "I" being "only a convenient term for somebody who has no real being," like Mary Hamilton, an identity that's really a rhetorical strategy used to voice taboo desires. "Lies will flow from my lips," Woolf-as-Beton says, "but there may perhaps be some truth mixed up with them; it is for you to seek out this truth and to decide whether any part of it is worth keeping."

In other words—or in words that describe how I actually used *A Room of One's Own* to construct my life as a writer—what's true is what's useful, and forget the rest.

It is possible that herbal-abortion recipes have been lost to the general public for a good reason. Compared with pharmaceuticals, they are not safe or reliable. Even *Eve's Herbs* is not an argument for restoring this knowledge so much as a correction to the assumptions that birth control didn't exist before the pill and that plants weren't used as drugs.

Just as it is possible that my great-grandmother did die from measles, not a back-alley abortion, or that my grandfather was better loved by his stepmother than the story he told. After his funeral, we found two pictures of my great-grandmother, one of Edith in her coffin in a sea of lilies and ferns, and one of her very alive, a dark-haired young woman dressed in men's clothing and grinning like a maniac, a costume and expression her great-granddaughter cannot, in the absence of letters and family lore, hope to interpret accurately. Nor can I hope to know what my grandfather was thinking when, thirty years after his mother's mysterious death, he brought two baby girls home—first Julie, then a year later Jennifer, both adoptions brokered by the obstetrician who delivered them of unwed mothers, a man my grandfather knew from his Catholic men's club.

It is possible that juniper is, most of all, a culinary spice whose virtues I have bypassed in favor of more dramatic themes, or that I could have spent this whole chapter talking about gin, once considered a feminine spirit, and still have found ample opportunity to discuss sexism and shame, the church lurking behind each turn of the page. How, in writing about abortion, did I barely mention the Catholic Church? Let us now revisit the divinity of juniper berries: they are, as the poet Melissa Kwasny writes, "hard, / dark, the nipples of some god we / dare not pluck."

"Gossip records a miracle," says Pliny the Elder, as recorded by John Riddle in *Eve's Herbs*, where I finally find something close to a juniper abortifacient recipe. To prevent conception, rub crushed juniper berries "all over the male part before coition," Riddle reports. "Another source near the same time had much nearly the same message, except that the directions were for the crushed berries to be placed on the vulva prior to insertion. One way or the other, juniper is inserted in the woman's vagina, and there it will act either as a contraceptive or an abortifacient, as Galen stated. Juniper was also taken orally." Good, finally, to know.

J's termination *was* my business, she said after reading a draft of this chapter. Her story was mine, too. "I invited you into it," she said. "I gave it to you." In the years since her procedure, she's been invited, as all women have, to shout her abortion on social media, a movement J admires but doesn't participate in. Too much exposure, she says. Her mother—she still doesn't know. "Even at Planned Parenthood," J says, "they asked me if I wanted to hear the heartbeat.

"As I asked for advice, I was so scared of judgment, though half the women I talked to—even the married moms—had abortions," she says. "I wish when I was making my decision I'd understood how normal this is. For women, abortion is the second-most-common medical procedure."

I'd heard this statistic before. When I looked it up to make sure it was true, I discovered that "abortion is the second-most-conducted surgical procedure in this country" is a baseless claim used by Rick Perry, former governor of Texas, to stoke antiabortion fervor in 2013, then alchemized by J a few years later into words that helped her make peace with her choice.

Both "medical procedure" and "surgical procedure" are misleading word choices. Medical procedures include everything from lab tests to major surgery; abortion is not a surgical procedure. The Poynter Institute's fact check of Perry's claim found that abortions are less prevalent than eye surgeries or intestinal surgeries or wisdom teeth extractions. Among reproductive-age women, the most common surgical procedure is the cesarian section; the second most common is the hysterectomy.

Regardless, J is right. Abortion *is* normal. If what is true today continues to be true, nearly one in four women will have one by the age of forty-five. No matter what words we use to describe normal, how we shelter abortion's history in lies, obfuscate our old medicines, or threaten, narrow, and revoke the legal rights of women to choose, no matter how much we feel and inflict shame—abortion is still normal. If the recipes for juniper, tansy, pennyroyal, and rue scattered in our archives tell us anything, they tell us this is true.

Juniper Jelly

The most appropriate recipe for this chapter is savin tea, but I never did find a text that could tell me how to make it, and, unlike the potential poison of "almond" extract I suggest risking in an earlier chapter, all my research suggests that making a savin-tree tea isn't worth the risk. So, instead, something sweet. Like jelly.

I wanted to make something that would concentrate and sweeten the medicinal zing of juniper, something one could eat with any kind of cheese, meat, or pickles. In this apple-based pre-serve, flecks of pulverized juniper float in an amber jelly. If you let at least a month or two pass, the pungency of the juniper will in-tensify. Pregnant women probably shouldn't eat this.

This recipe makes two batches of apple jelly, or about 56 ounces. It's based on apple-jelly recipes from *Mes Confitures* by Christine Ferber and *Saving the Season* by Kevin West. West talks about apple jelly as a "blank slate" that carries other flavors, which I've interpreted by adding a prodigious amount of juniper. This recipe needs a mortar and pestle.

Yield: about 56 ounces

2 kilograms (about 4½ pounds) apples, quince, and/or crabapples (whatever you have)

2 kilograms (about 8½ cups) water

6 cups sugar, divided

4 tablespoons lemon juice, divided

60 freshly crushed (and ground to smithereens) juniper berries, divided

Cut the fruit into quarters; cut off the stems and calyxes, but leave the peels, cores, and seeds. Cut the quarters into quar-ters again. In a stockpot, combine the fruit and water and sim-mer, partially covered, for 30 minutes without stirring. Strain through a damp jelly bag or a double layer of cheesecloth into a bowl. Let drip for 30 minutes, then squeeze the bag gently for the last drips of pectin (don't squeeze hard—it might make the jelly cloudy). Discard the fruit. Store the fruit-pectin stock in the fridge overnight to let the solids settle to the bottom. It will keep there for up to a week.

Prepare a deep canning pot with boiling water—enough to cover eight 4-ounce jars. I add a little white vinegar, to mitigate the powdery white residue my hard water leaves on my jars after I boil them. Sterilize the jars by keeping them immersed in the boiling water for 10 minutes, then set them on a clean towel to cool and dry. Boil the lids for 10 minutes, then place them on a clean, lint-free towel, seal-side up, to dry. Rinse the bands and set them aside. Keep heating the pot of water on low until the end of this recipe, when you'll use this water bath to process the filled jars.

Measure 3 cups sugar and set it aside for use in the second batch of jelly. Then put 4 cups fruit-pectin stock (taken from the top of the bowl, leaving behind any solids at the bottom) and 2 tablespoons lemon juice into a deep-sided preserving pan (a heavy Dutch oven works well). Boil over high heat for 5 minutes without stirring, starting the timer once the pan is at a full boil. Add the remaining 3 cups sugar, stir until dissolved, and boil over high heat for 7 or 8 minutes, stirring the bubbling foam at the top to keep it from boiling over, until the jelly darkens a little and doesn't quite coat the back of a spoon. If at any point you can't control the boil by stirring, turn the heat down before the jelly boils over. Add half the ground-to-smithereens juniper berries to the jelly, and boil for another 1 or 2 minutes or longer, until the jelly sets. Test the set by covering a wooden spoon in the jelly and setting it aside on a plate for 1 minute (take the jelly off the boil while you do this). If the jelly wrinkles when you return and push your finger through it, and if it gathers at the edge of the spoon before dripping viscously, it's done.

If there is scum on top of the jelly, spoon it off, then ladle the jelly into the 4-ounce jars, leaving ½ inch of headspace. Wipe the rims clean, place the lids, and screw the bands on, fingertip-tight (tight, but not so tight you need strong hands to reopen them). Bring the water bath back to a boil and process the jars, keeping them immersed for 10 minutes. Remove the jars from the water, and set them on the counter to cool.

Don't worry if the jelly appears very liquid. It will set as it cools. It will set even more over time. Store at room temperature in a dark cupboard, and refrigerate after opening.

This recipe makes about 8 cups pectin stock, so you can come back another day that week to make another batch with more or less juniper, to taste, following this recipe again from "Prepare a deep canning pot with boiling water."

Juniper Bitters

Bitters are aromatic extractions of roots, herbs, and spices. Neutral grain spirits wring every drop of flavor from plant material and usually much of the plant's color, creating a rainbow of essences that are bitter or sweet or funky or herby. Bitters can be made from a limitless selection of plants; if this is your first time making them, don't stop with this recipe. I particularly love bay, star anise, yarrow, and mugwort bitters for adding unusual flavors to jams, pies, drinks, and other concoctions.

My version of juniper bitters combines the ginny zing of juniper with the citrus of lemon zest and the bitterness of lemon albedo, plus rosemary, which, when extracted, adds a surprising subtle sweetness. Black and pink peppercorns and coriander round out the flavors. The juniper can get a bit buried under the lemon and other botanicals; pump that flavor back up by combining the juniper/lemon/rosemary bitters with pure juniper bitters.

This is probably not savin tea, which I'm guessing (since my sources don't provide recipes) is a decoction made with water, not alcohol. But this extraction would be just as powerful, if not more so. Use these bitters sparingly—in drops, not drams—as you would any bitters.

Yield: about 5 ounces

6 tablespoons juniper berries (fresh or dried), divided

2 pink peppercorns

2 black peppercorns

2 coriander seeds

Peel of ⅛ lemon, pith intact

Rosemary sprig

7 tablespoons Everclear (buy a whole bottle—it's cheap, and you can use the rest to make other bitters)

With the side of a chef's knife, lightly crush 3 tablespoons of the juniper, all the peppercorns, and all the coriander. Add them to a small jar. Add the lemon peel and rosemary to the jar. Pour Everclear over the botanicals until they're covered, about 4 tablespoons.

Lightly crush the other 3 tablespoons of juniper, add them to a separate small jar, and pour Everclear over them to cover, about 3 tablespoons.

Cover the two jars and let them sit for 10 days.

Taste the contents of each jar to make sure the bitters are as intense as you'd like them to be. Then strain the solids from the bitters. Discard the solids. Combine the bitters in a small bottle. Store at room temperature.

K: Kiwifruit

Actinidia deliciosa
Actinidiaceae (Chinese gooseberry) family
Also known as kiwi, Chinese gooseberry, yang tao

The June I was nineteen, I took a summer job as payroll officer for the same nursing home where my mother worked, and I learned not to mind the shit smell beneath the bleach smell or be too bothered by my own mortality, how the evidence before me confirmed what I already knew from years of visiting Mom's facilities: with the right combination of senility, injury, poverty, and will to live, I, too, might end up like Lucille. Every day, she wheeled herself down a hall of open doors, hunched in the shape her spine now assumed, her skinny arms like downed limbs in her lap, mumbling, "Help me, help me, help me."

"She's fine. Don't help her," Marla told me my first day at work. "And don't let her out of the building."

My qualifications for payroll officer at Pleasant View were as follows: I had a high school degree, my mother was one of the bosses, and Accounting needed someone who wouldn't mind being laid off in September. I worked with Candy in Reception, Marla in Accounting.

Candy was old Portland and looked like a shopping-mall ice-rink champ. When she answered the phone, her smoker's growl was taffy-sweet. Marla was from Astoria, had a ladybug tattoo on the web of skin between her thumb and forefinger, and had three grandchildren who lived with her sometimes. If the babies came to her for comfort in the night, she didn't want to scare them with her sleep-apnea machine, so she'd painted the mask and its thick plastic oxygen tube to look like an elephant. She taught me the computer systems, didn't get mad when I made stupid assumptions about people or filing jobs, and shielded me from the certified nursing assistants whose paychecks I messed up the first time I processed payroll. "If you learn from your mistakes," Marla promised, "they'll like you again."

When I wasn't processing paychecks, my job was data entry. Half the work was strategizing how to stay awake while maintaining the appearance of diligence and Marla's good opinion. It was easy to get sloppy with the [insert name, return, code, return] of entering a new patient into the system or [delete, return] of taking one out. To stay sharp, I collected first names in a notebook, as if I could make a complete set: Mildred and Beatrice, Norma and Ernest, Wilbur and Barney and Opal. I imagined their equivalents in people my age, how those names would sound in 2052. Time might make "Megan" and "Michelle" quaint, or "Crystal" and "Bryce" beautiful, or "Zach" and "Jordan" as strange as "Hortense" and "Hellmuth." To delete a name meant the patient had recovered, transferred to a different facility, or died. My favorite name was Beulah. I thought it meant "heaven." I entered Beulah in the system and took her out without ever having met her.

There were early mornings after late nights when I'd wedge my body behind the filing cabinet and sleep until the minute before Marla arrived. Sometimes I'd calculate payroll with a cockatiel named Judy perched on my shoulder, shitting down my back. Birds like Judy were supposed to help the residents. The staff loved having a small creature whose grip on our shoulders and nibbling of our earrings broke up the repetition of the day, the breakfast/lunch/dinner/bath rituals of getting well, or getting further toward dying.

People often left the nursing home healed. That summer, I saw their triumphs briefly, through the blinds of the window I'd just slept under, as their families held the door and wheeled them over the threshold. But the success stories weren't the ones that reached the office on long rumor trails.

There was comatose Kim, the victim of a car accident, T-boned on her way to the beach six years earlier. The story went that she had a bad relationship with her mother, but now her mother was there every day to turn her, feed her, talk to her. "You see," my own mother said, "why it's my job to worry about you."

There was the evil doctor who hid his money when he

was well so that when he was sick the state would cover his care. While he lived at the home, he married a fellow resident, Betty. Mom said that she'd once had to stop him from running Betty down with his wheelchair, that he called his wife nasty names in front of the staff. The thing was—so the story went—Betty didn't mind. She had Alzheimer's. She remembered she was in love, not that her lover was an asshole.

Anastasia would sit cross-legged on the floor by the front door, not a flight risk, just hoping someone would take her back to Czechoslovakia. Gordon stopped playing it straight and started hitting on the male CNAs. Mary had always been reclusive: "Get out of my house!" is what she'd yelled at passersby before blasting them with her garden hose. "Get out of my house!" is what she yelled at CNAs who now darkened her door with meal trays.

There was Annabeth, my favorite, a former silent-movie pianist who could play "Bye Bye Blackbird," "Oh, You Beautiful Doll," and other classics if you asked for them. As her dementia progressed, "You Beautiful Doll" ceased to be a song and became a name she called me. She forgot key signatures before forgetting "Bye Bye Blackbird," playing one hand in C and the other in E. For a while she would play if you sat her at the piano and put her hands to the keys. Then she forgot that. Then she forgot how to swallow. Then she died.

Peacefully, my mother said.

Middays, I'd meet Mom for lunch if she could clear her schedule. We'd take our brown bags into the back parking lot, where no one would bug us, eat PB&Js while sitting on

concrete tire-stops with a view of nothing pleasant or unpleasant, just the vinyl siding of the neighborhood behind
the nursing home. The contents of these lunches hadn't
changed since my childhood. Sandwich, carrot sticks,
crackers, fruit, the food groups of a working mom. I can't
remember now if I made brown-bag lunches by her example that summer or if she made them out of habit. Maybe
we traded off. I do remember that Mom still had a way with
lunch fruits. She'd cleave an apple from its core or denude
a kiwifruit with the swift precision other mothers reserved
for eye makeup. With a paring knife in her right hand and
a kiwi in her left, she'd cut a flap from the rough brown
peel and press that flap between her thumb and the flat of
the blade. Then she'd strip the kiwi down, turning it on its
axis until she held a slick green globe of fruit, which she'd
finish by slicing it into cheerful medallions, each white
center haloed with black seeds. She had big hands like her
father's, she said, shy about them even though they made
cutting fruit look like ballet. She was shy about her teeth,
too, which were slightly crossed in front and a little gapped
near the gumline. When she took a big bite of jam-on-toast,
jam would squeeze out between those front teeth, a detail
I loved but knew she wouldn't want me to notice. I have
my own way of slicing apples now (quartered, laid flat on
their sides, the core removed in one angled cut from each
quarter), but I no longer eat kiwifruit. For no good reason
except that I'm too old to have my mother cut fruit for me,
and there's no other way I want to eat a kiwi.

Mom ran the physical-therapy department at Pleasant
View, which meant that sometimes at lunch she'd still be
with a patient, walking behind him as he crawled along

parallel bars, holding him by a hip harness while he practiced keeping his balance. One man couldn't walk a step though there was nothing physically wrong with him. Psychosomatic, Mom said. During therapy, she'd work his feet, stretch his ankles, his legs, move the joints to keep them loose. Once, she did get him to walk. The moment she pointed it out, he collapsed.

At the end of the day, when I was done with my data and filing, done with the birds and phones, I'd walk the length of Pleasant View to Mom's office, at the opposite end of the building. She'd be at her desk, glasses on, paperwork and pen flying, headache flushing her nose and forehead and chin, telling me, "Just five more minutes, honey, I'm sorry, I have to finish these notes." We'd make bargains to get her out of work on time. "Be tough," she said. "Make me go, or I'll stay all night." Then she'd say, "Just one more thing."

This is why she's always late, I realized. Once her work is done, she finds more work.

Even now, I catch myself saying what I heard her sigh each evening that summer: "I didn't get enough done." As if we expect to encounter a task whose completion will flood us with satisfaction or pride or peace, about which we can say, "There, I've done it!" and put down our paperwork.

In the nursing home, "enough" was a decision. I'd decide with my impatience, spinning in an office chair until her guilt at staying exceeded her guilt at leaving. Enough, I said. Let's go home.

When I was a child, I thought of my mother's profession in terms of how my brother and I rose earlier and came home later than other children, our days bookended by day care, and how she knew everything about how to help us feel better—how to ease a sleep-kinked neck, how to soothe a sunburn, when to cut a high fever with a cold bath, the broccoli we must eat. Her workplaces were large rooms full of exercise equipment, stiff beds whose thin comfort only amplified a child's restlessness, elastic ligatures and giant therapy balls that could have been toys but were too awkward to entertain without mischief. People came to her offices to learn minute, complex, and repetitive motions that somehow made them well, exercises she often taught through the static of a migraine.

My brother and I knew she had headaches, but on the days when she got out of bed and went to work anyway—most days—we'd forget. Sometimes, despite her efforts not to show how she felt, she'd slip. Like that day when we would not stop bickering and, rather than leave the house to run the gauntlet of our after-school lessons, our mother dropped her briefcase, sat down on the floor, and wept.

Years after my job at Pleasant View, my mother would confess she wished she hadn't worked full-time while we were kids. I tried to tell her how she was wrong. "But you were a great example to us," I said, "to see you passionate about your work." I loved her work the way I loved her fruit. I loved the fact of them, how I could take for granted that my mother loved her job and was good at it. How, in our house, ambition and caretaking were often the same thing.

"But I was so sick then, working like that," she said. "And I didn't have to be."

Near the end of my job, in early September, I stepped out of my office to get a midday Coke and found Lucille and her wheelchair blocking the front door. She had helped herself all the way down the hall. Once she'd gotten there, she'd reached out her arm, palmed the bar that would unlatch the door, and right there, within that gesture, right before she triggered the alarm, fallen asleep.

"I told you she didn't need help," Marla said.

I left Lucille alone and took the long way to the Coke machine, a direct route through the bustle of lunch service. Patients were propped up in bed with cafeteria trays in their laps. Patients filled the dining hall with their wheelchairs and walkers. CNAs swerved around me with platters of food, the smell of butter and corn and roast beef stronger for the moment than every other human smell. Past that were more patient rooms, then treatment rooms, then Mom's office, where I would collect her from her reports and insurance claims later that day. On my way back with the Coke, I caught a glimpse of her kneeling in front of a retired English teacher named Rose, a sweet lady who was "still with-it," we'd say, with the overconfidence of the ablebodied. She'd been living at Pleasant View while recovering from a broken hip. My mother was kneeling, looking at Rose, narrating how these exercises would help her relearn her body, naming each way Mom would support her as Rose gripped parallel bars and attempted to walk three feet. In a couple weeks, this place would once again be only my mother's, its cycles barely ruffled by that kid in the main office. She'd go back to finding her own way of saying "enough" and calling the day done, or saying "enough" and just calling it a day.

I waved to her. I practiced getting back to work.

The Perfect Peel

Kiwis, the furry flightless birds kiwifruit was renamed for, were called kiwi by the Maori for the sound of the male bird's mating cry. Native to New Zealand, one of kiwifruit's main producers, kiwi birds are the rare species whose female is larger and more aggressive than the male. When a suitor irritates her, she might kick him away. Kiwi eggs are large in proportion to the female kiwi's body. By the end of gestation the egg overtakes her, leaving too little room for food and making movement a painful, awkward, waddling chore. Once she lays her egg, her male partner—who bonds with her for life—does most of the nest sitting.

Kiwi*fruit*, a sweet, green, ovoid fruit a bit larger than a chicken's egg, originated in China and had several names in the twentieth century, the most recognizable of which to English speakers is "Chinese gooseberry." Translations of the Chinese names for the fruit include "monkey peach," "goat peach," and "goose fruit." In French, the first name for kiwifruit translates to "vegetable mice." Kiwifruit or just "kiwi" appeared in English in the 1950s, renamed from "Chinese gooseberry" by growers who wanted to prevent confusion at the American border, where customs officials were at the time enforcing bans against certain fruits from communist countries, including China. Kiwi is now used as an affectionate term for a person from New Zealand, too.

Known for its beautiful green, chlorophyll-packed fruit and a corona of black seeds, kiwifruit is almost always eaten fresh. Thanks to actinidin, a naturally occurring enzyme in kiwi flesh, mashed up raw kiwi can be used as a meat tenderizer or to speed the ripening of other fruits. Raw kiwi will turn milk and keep gelatin from setting.

How to peel a kiwi is a minor point, and one I only paid attention to because my mother was so graceful at it. Sam, seeing me trying to peel the kiwi the way she did, in longitudinal sweeps from top to bottom, says he peeled kiwi for his kids when they were young by slicing the fruit in half, then scooping it from the peel with a spoon. Or slicing the kiwi in half and handing one half and a spoon to each kid. One can also just slice the kiwi in thin rounds and eat the peel. Or eat them whole. The peel is fuzzy, edible, and not unpleasant.

Meat Tenderizer

Kiwifruit offers a quick trick to anyone who didn't get home from work in time to marinate meat for dinner. Actinidin, the enzyme in kiwi that dissolves collagen, is a particularly effective tenderizer because it doesn't turn meat to mush, as the bromelain enzyme in pineapple can do if not used with care—though actinidin will liquefy meat if allowed to work for hours.

A little goes a long way. Some cooks recommend adding 1 to 2 tablespoons of kiwifruit purée per 1 cup of marinade, then spreading the marinade over meat for 10 to 15 minutes before cooking. Others recommend using no more than half a peeled kiwi for 4 to 5 pounds of meat. Mash the kiwi in your hand, spread it over the surface of the meat, and leave it to marinate for 10 to 15 minutes. Then wipe the kiwi off, season the meat with your preferred mix of salts and spices, and cook.

A study published in the medical journal *Burns* suggests a fascinating use for kiwifruit as a treatment for debriding damaged skin. In their experiment, kiwifruit actinidin gently broke down devitalized tissue, making it easier to remove and clearing the way for faster healing. What works for dinner works for wounds, too.

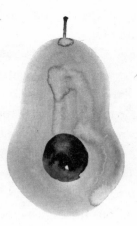

L: Lump

"It isn't fair," we said when L's beer belly turned out to be a tumor. "Not fair," we said when L lost her hair, her ovary, love interests, and friends who couldn't cope. We were one month into our last year of college when cancer crashed our party. No one said—no one knew how to say—what was also true: what's more fair than cancer?

Fifteen years later, I had my own scare. It arrived while I was writing the first draft of this book, the week of Christmas, three months before my first deadline, with half the chapters still to write. I never check my breasts for lumps, but that day I checked. On the right, nothing but the usual flesh. On the left, nothing.

Then something. A firm spot I could find if I put my fingers in my armpit and dragged them toward my nipple. For the next two days, I checked it in secret, slipping my hand into my shirt in the bathroom, in the car, rooting at my breast, afraid to scare Sam, still not sure I was feeling what I was feeling. The lump was the size of a blueberry. My doctor said chickpea, but I wanted a fruit comparison, like pregnant women get for their fetuses. First a berry, then a plum, then a grapefruit, then a melon. Then you have a baby or die.

In an essay I wrote the year after L's diagnosis and never let anyone read, I describe her smiling as she came home from the doctor's office, pointing at her distended stomach, the belly we thought she'd made with beer. "It's a two-year tumor!" she crowed. An ovarian cyst had grown into tissue that looked so maternal, the student health center initially thought she was six months pregnant. They assumed the pain that brought her to the doctor was her body spontaneously aborting a fetus. Leave it to a university campus clinic to think a twenty-two-pound tumor was a baby. At least they didn't think it was an STD, I said.

Maybe I said that. I had a problem with humor while L was sick. I couldn't get the hang of how it worked. When she said her paunch was a two-year tumor, that was funny. When I said we should name it Watson, that was not funny. When her hair came back, she told anyone who liked her haircut that it wasn't a haircut, "it was a hair*grow*," and I said she looked amazing, "never should have had long hair," like it was lemonade she'd squeezed from that lemon of a year.

Her eyes went glassy; she wouldn't look at me. I shriveled with shame.

The rule was simple. She could joke. I could laugh but not joke.

Still, I joked.

Between chemotherapy infusions, she crashed on my couch. I'd throw myself into preserving the shreds of our normal college lives, rock shows and parties and movie nights in, a frantic fun that wasn't fun but was better than staying still while she waited for health and hair to return. She was going to get a tattoo over her missing ovary. "Maybe a heart or a star," she said. "Something purple. I can't have children, so I won't have to worry about stretching." In the essay from the year after cancer, I wrote that L could "pass" for healthy, her bald head camouflaged in a beanie that made strangers think she'd shaved her long brown locks for kicks, not sickness. "I LOVE YOUR HAIR!" an acquaintance shouted over loud guitars and drums. "SO PUNK!!"

"THANKS!" L shouted back. "I HAVE CANCER!"

Watching that girl's face fall, feeling my own guts churn (with what: Worry for L? Sympathy for the girl?), I wondered if on some level L enjoyed this. If she was going to be sick, she was going to shock you with it, the power of her confession meted out and controlled with party-trick precision, a performance of coping that blew us all away. I wondered this and said nothing and buried the thought, that slimy seed of resentment. I felt like cancer's sidekick, and I hated it.

My mother and her six siblings have had cancer scares and actual cancer too many times to count, traded in mutinous organs for more years, consented to dozens of painful procedures to determine that the lingonberry- or longan- or lychee-sized lump they'd found wasn't going to kill them after all. Only after my latest mammogram came back negative did I confess my terror to my mother, a tearful apology that started with, "I've had my first cancer scare." *First*, because family history says I can assume this will happen again. *First*, because in our family, as in so many families, cancer scares are a stage of development, like first periods and first babies and male pattern baldness. I'd joined my mother where she'd been living for decades, as a woman whose body could ripen the wrong way. Saying it out loud was a relief. It felt good—can I say that?—to step into that garden, to find a comfortable seat. To have and be her company.

When I was twenty-one, I wrote that the misdiagnosis of L's tumor was "enough to stop her heart for a second. The nurse: 'Surprise! You're pregnant!' so the first question she asked herself was not *why me? why now?* but *what will I name it?*"

The late summer before our cancer fall began, my boyfriend dumped me, L didn't get a role in a play she'd auditioned for, and instead of moping in our basement apartment we said fuck it, we have time, we have gas money, let's drive to San Francisco. Maybe I remember this wrong. Maybe L just saw I was sad. Either way, seventeen hours later we drove over the Golden Gate Bridge, and a few hours after that I took a photo of us at Fisherman's Wharf. The sun's all over L as she leans back, soaking it up, beautiful and free in this brief exit from our lives. Her shirt rides up so you can see her belly, the firm round curve of what we don't yet know is cancer. In this photo, it's cute. We're so young.

Who hasn't had a cancer scare? Or cancer? Or loved someone who has? Whose body won't suddenly—or slowly, over the course of several years, secretly—bear an orchard of deadly fruit? Who has stopped themselves from harvesting a plot from this mess by making cancer a test of character or a judgment of God? One of those illnesses Sontag wrote about, like TB or AIDS, whose mysterious cause and mysterious cure make them ripe for magical thinking.

I no longer speak to L, so I can't fact-check whether or not her tumor really was twenty-two pounds, though everything I wrote at the time indicates that it was.

That's the good news—L survived! And had children! The only thing I can confirm that cancer killed, besides some of L's memory, was our friendship. We held on for a few years after, like a peach the wasps ate, skin and pit dangling from the branch but all the soft parts gone. She would have kept going, I think. But I couldn't recover. I was too tired, too bitter, too consumed with her survival. She was well again—what a relief. She was well again, and all I wanted was to get away from her. Cancer I could cope with, even with my foot in my mouth. Recovery—that's when I really stopped knowing what to say.

After I found a blueberry-sized lump in my left breast, after a week of secret panic, after taking off my bra and putting on the hospital smock, after that lump showed up on a mammogram and ultrasound screen as a common cyst—three of them, actually, but nothing to be afraid of—I put on my shirt and called my mother. I asked her if a lifetime of false alarms made her regret her vigilance, her batteries of tests, her alkaline diets, her ovariohysterectomy, her fear. She said she didn't know. She said it was impossible to know.

Recipe for a Long Lost Friend

Text collaged from John George Hohman's Pow-Wows; or, Long Lost Friend: A Collection of Mysterious and Invaluable Arts and Remedies, Good for Man and Beast, *the "famous Witchbook of the Pennsylvania Dutch," 1820.*

Those who neglect the use of this book to heal the sick may forfeit all hope of salvation. For all this is done by the Lord, who makes sickness an opportunity of faith. That thee might be cured is a sign He verily exists. If thee does not call on Him as He commands, this will be an abuse of illness.

Remember at such times His remedies:
To prevent Death, seek refuge beneath the tree of life, which bears twelvefold fruits.

To prevent Wasting of the Body, name thy lumps for small fruit and say their pits are seeds of holy trees.

To prevent Hysterics and Colds, take off your shoes and socks, run a finger between each toe, then smell it. Do this every night.

Be willing that thee should be seen by all women and men as you confess to them, and let His arms wrap your body in comfort as the body wraps its lumps in blood.

M: Medlar

Mespilus germanica
Rosaceae (rose) family
Also known as cul de chien, open-arse

A shriveled, rose-hip-like bulb about the size of a fig, called in Shakespeare's time "open-arse" because its calyx looks (sort of) like the pucker of an anus. Called open-arse or cul de chien today because once one gets a nickname like "open-arse" or "dog's ass," even centuries can't erase it. That medlars are not ripe until they rot—a process called bletting—contributes to their assness, I guess.

When Shakespeare was staging plays, medlars were the last ripe thing on the branches before winter blasted in, the only thing besides honey that would have sweetened a not-extremely-wealthy-person's bread.

Not wanted, not exactly. Wanted when there was little chance of any other sweetness.

Mercutio re Romeo, when Romeo is pining for Juliet and hiding from his friends: "Now will he sit under a med-lar tree / And wish his mistress were that kind of fruit / As maids call medlars when they laugh alone," Mercutio teases. "O Romeo, that she were! Oh, that she were / An open arse, and thou a poperin pear."

As far as I can tell, no one cultivates medlar commer-cially in the United States, though maybe someone has a tree on their property and brings the fruit to a farmers' market. This would have to be a place where farmers' mar-kets convene in late fall. California or Seattle, maybe. But not Spokane.

I find medlars through two sorts of people: arborists who plant rare fruit in their home gardens, and artists who maintain mental maps of where to go strange-fruit scav-enging. A neighbor's side yard. A city arboretum.

When I luck into a few pounds of medlar (my luck is never larger than a few pounds), I rediscover that they are one of the world's more patient fruits, able to lie in the basement without complaint, bletting slowly on baking trays from mid-October to early December, while I decide how best to cook them. They should be arranged in a sin-gle layer, some say on straw (though I don't find this nec-essary), then left alone. Over several weeks, they transform from a hard gold to a mushy brown. When you can split them open with a firm squeeze, they've bletted.

In Shakespeare, medlars are not a metaphor for a woman no one wants. I'd assumed they would be, these fruits that are rotten when they are ripe. They are instead what a jackanapes like Mercutio calls the woman who competes for their best friend's attention, or the woman they've escaped, so far, from marrying, not because the woman is unmarriageable but because the man is a bounder. It's a dirty joke, usually, about private parts.

So medlar is a metaphor for pussy or ass. And ass, if you're getting some, is a metaphor for pussy, which you can also get, because both just mean sex. This is basic metonymy.

Lucio to the Duke, to describe a woman he slept with and abandoned in *Measure for Measure*: "Yes, marry, did I but I was fain to forswear it; they would else have married me to the rotten medlar."

Lucio's medlar-woman is too accessible, too sexual, moist, effervescent, fermented. Medlar is, in Shakespeare, a way to call someone a slut. An insult that betrays, as name-calling often does, the speaker's insecurity.

Shakespeare refers to medlar three times in all his plays. I want to use medlar for all sorts of purposes, none of them edible:

1. As a metaphor for the woman no one wants because she is too complicated, not sweet enough, too old.
2. As a foil for market-made expectations that fruit should be sweet, and parallel expectations that *women* should be sweet.
3. Sweetness: a feminine mask. Worn to protect or rob or fool. Or for fun. Or for beauty.

The medlar tree itself has beautiful black curving boughs and soft white blossoms, the kind I have to hold myself back from joyfully crushing. At the end of the season, medlar fruits hang from the tree's bare branches like blackened Christmas ornaments.

Some advice has one pick medlars before the first frost, when they are bronze-skinned and hard-fleshed, store them in a cool place with a light source, and hope time will soften them to edibility. Other advice insists medlars must be picked *after* the first frost, so that plunging temperatures can soften their pectin-hard flesh and sweeten it, a quality they share with persimmons. Bletting, indoor or out, occurs unevenly throughout the crop. It starts at each core with a brown spot that spreads through the fruit until the medlar's skin darkens to a walnut color and its flesh yields to touch, something that if dropped would not bounce. Were one to cut a bletted medlar in two or squeeze the fruit crudely from its skin, one would observe that its insides are a brown pulp that cradles four or five whitish seeds about the size and shape of dried hominy.

Medlars that have been bletted indoors have a starchy, almost waxy texture that tastes of fermented apples. It reminds me of flavored lip gloss, the kind I don't mean to eat but do, just a little, every time I lick my lips. It might not remind anyone else of that. A medlar that has been bletted indoors has been sheltered from the wild rot that would improve its flavor, but, still, I prefer my medlars this way. They are more dependable than their outdoor-bletted counterparts, which are subject to poaching from winter-panicked squirrels.

A single medlar that has been bletted outdoors through early December can be eaten in three bites. The first taste will be of spiced applesauce. If its skin were sturdy, its flesh could be spooned from it like too-thick pudding. But it is not sturdy. Don't eat the skin—it isn't tasty. Don't eat the seeds, either (also not tasty).

The second taste, because the medlar has spent long cold weeks on the branch, is sparkling wine. Not a good sparkling wine, but pleasant enough. Slightly explosive-tasting, like certain manufactured candies. Ugly, but what a personality.

The third taste is a cold mildew one usually only smells, and generally interprets as a warning not to eat any more.

You have now finished the medlar.

Maybe your medlars will be bigger than mine. Maybe yours will require four or five bites. I'd say how nice for you, but I'm not convinced more bites will improve one's experience of raw medlar.

Medlar jelly, however, prepared by boiling bletted medlars in water to extract their essence ("juice," in this case, isn't the right term), makes a strange and lovely preserve that's hard to describe except to say it's sweetly woody, like rooibos tea, labor-intensive to prepare but extremely rewarding to eat, a specialty worthy of accessorizing with the best cheese.

What is the right virtue of the medlar? In *As You Like It*, the fool Touchstone says, "Truly the tree yields bad fruit." Rosalind, our charismatic cross-dressing heroine, not the un-lined offstage Rosaline from *Romeo and Juliet*, replies, "I'll graft it with you, and then I shall graft it with a medlar. Then it will be the earliest fruit i' th' country, for you'll be rotten ere you be half ripe." Unexpected, raunchy, sweet, and rotten—"that's the right virtue of the medlar."

Medlar Jelly

Medlars are pasty and pulpy, not juicy, but one can make an excellent jelly from a decoction of them. It will taste how boiling medlars smell—which sounds like something witchy, but is really more like roses, strawberries, apricots, and wine. The finished jelly has a sweet woody taste, slightly spiced. This jelly doesn't always set, particularly if it's made exclusively with fully bletted medlars, which have less pectin. If you end up with a syrup, don't despair. Pour it over chèvre and serve it like that's how you meant to make it.

Depending on when you pick them and where you store them, medlars take weeks to start bletting and continue to blet for several weeks, after which they begin to rot in earnest. I received my medlars on October 15, they began bletting just after Halloween, and I made the last batch of medlar jelly on November 30. By then, a few of the medlars had gone bad. What is the difference between a well-bletted medlar and a bad medlar? A bletted medlar will be nut brown, soft, smooshy, a brown smear; a bad medlar will be dry and black. Partially bletted medlars will be a rich maple color, and will blet from the inside out. Cut one open and you'll see a russet spot beginning right at the center and spreading from there. When medlars go bad, they start molding from the inside.

To boost the pectin content of the jelly, combine bletted medlars with a cheesecloth bag of pith and seeds from yuzu or the squeezed-out, zested remains of conventional lemons. I used 2 cups of pith and seeds (from four yuzu) to test this recipe. If that option feels too fussy, add some partially bletted medlars instead. Over time, I find that medlar jelly thickens up much more than other jellies, and lemon pectin seems to create the best long-term texture.

I've also found that, after a year in storage, sugar crystals start to form in medlar jelly. In general, this means a jelly has either been cooked too long or too little, but in this case, since my jelly method hasn't produced crystals in any other fruit jelly, I think it may have something to do with medlar itself. It's still a mystery to me. In any case, if you happen to have medlar jelly sitting around a year later, fish out the crystals and eat them. They're delicious *and* pretty.

If you don't have a vanilla bean or don't want to buy one, don't substitute vanilla extract; it masks the unique flavors of the medlar. Just leave the vanilla out.

Yield: 32–40 ounces

1 kilogram (2.2 pounds) medlars, mostly well bletted

6 cups water (or enough water to cover)

Sugar

A cheesecloth bag of lemon skeletons and seeds, 1 or 2 cups (however much you have on hand)

Juice of ½ lemon

½ vanilla bean, slit open along one side (optional)

Cut each medlar in half and place—skin, seeds, and all—in a preserving pan. Cover with the water, then boil gently, with the lid slightly ajar, for 1 hour. Strain the pulp through a damp jelly bag and let the juice drip for 2 or 3 hours into a bowl, massaging the jelly bag gently at the end to coax more pectin from the fruit. Do not squeeze hard—that will make your jelly cloudy. Cool the juice in the refrigerator overnight to let sediment settle to the bottom. Discard the medlar pulp, or reboil it for medlar syrup (following recipe).

The next day, prepare a deep canning pot with enough boiling water to cover five 8-ounce jars. I add a little white vinegar, to mitigate the powdery white residue my hard water leaves on my jars after I boil them. Sterilize the jars by keeping them immersed in the boiling water for 10 minutes, then set them on a clean towel to cool and dry. Boil the lids for 10 minutes, then place them on a clean, lint-free towel, seal-side up, to dry. Rinse the bands and set them aside. Keep heating the pot of water on low until the end of this recipe, when you'll use this water bath to process the filled jars.

Measure the juice (I extracted about 3 cups), leaving behind any sediment that has settled at the bottom of the bowl. Measure out the same volume of sugar (in my case, 3 cups) and set it aside.

Wet some cheesecloth or a jelly bag, fill it with the lemon skeletons and seeds, and tie it shut. In a high-sided preserving pan (like a Dutch oven), boil the juice and the bag of pith over high heat for 5 minutes, starting the timer when the juice reaches a rolling boil. The lemon bag will not be completely submerged. Remove the pan from the heat, milk the bag of more pectin by holding the top with tongs and squeezing another pair of tongs down the bag. Discard the contents of the cheesecloth bag.

Return the preserving pan to high heat, and add the sugar and lemon juice, stirring to dissolve the sugar, and bring the pan back to a boil. Add the vanilla bean, if using. Keep boiling over high heat (I mean it—high heat), allowing the medlar syrup and sugar to bubble up the sides of the pot, stirring only to keep the jelly from boiling over, until it's set, about 3 to 5 minutes (though this once took only 90 seconds, so keep an eye on it). After a minute of boiling, run a wooden spoon through the jelly, hold it up, and see if the jelly congeals a little at the edge of the spoon before dropping off. Repeat every minute or so. Once this congealing occurs, remove the jelly from the heat.

Now test the set: Cover a wooden spoon in the jelly and set it aside on a plate for 1 minute (take the jelly off the boil while you do this). If the jelly wrinkles when you return and push your finger through it, and if it gathers at the edge of the spoon before dripping viscously, it's done.

Remove the vanilla bean (it can be rinsed off and reserved for another purpose). If there is scum on top of the jelly, spoon it off. Ladle the jelly into the prepared jars, leaving ½ inch of head-space. Wipe the rims clean, place the lids, and screw the bands on, fingertip-tight (tight, but not so tight you need strong hands to reopen them). Bring the water bath back to a boil and process the jars, keeping them immersed for 10 minutes. Remove the jars from the water, and set them on the counter to cool.

Don't worry if the jelly appears very liquid. It will set as it cools. Over time, it will set even more. The jelly may taste strongly of lemon in the first several days. The flavor will settle down over time, and medlar will reassert itself.

Store at room temperature in a dark cupboard, and refrigerate after opening.

Medlar Syrup

I tried so hard to find a delicious use for medlar pulp.

I failed.

Medlar jam is a boring brown paste with a predictable apple flavor. Medlar tarts, same. Medlar ice cream was somewhat better, but basically—same. Medlar liqueur is tart and astringent, with that same so-so apple flavor and none of the delicacy of medlar jelly. Medlar bitters and medlar tea were not worth the trouble. Medlar vinegar and medlar catsup were fine, but not better than more common vinegars and catsups.

My problem may be my medlars. They're indoor-bletted, with a starchy flavor and texture. Outdoor-bletted medlars are sweeter, with less detectable starch, but I never find enough of these to make a full batch of jelly. My main medlar supplier (my friend Ellen) lives on the other side of the Cascade Mountains and must pick and send medlars while they're still hard so they can make the journey intact; my local medlar supplier (my friend Kelly) harvests only a handful of medlars a year because of the youth of his tree and medlar's popularity as winter food for foraging animals. If you live close to a large medlar tree (or have one!), try to obtain a large quantity of outdoor-bletted medlars and let me know if your experience with their pulp is delicious.

Either way, medlars' best flavors are delicate. They are very much like a sweet, spicy whiff of wet autumn leaves, the sort of perfume you walk through while entering a warm house on a cold afternoon. These flavors are only apparent in the liquid created by boiling medlars. Even that flavor is drowned when combined with anything else. Even soda water!

And so. To my taste, the best culinary use of medlar is as syrup or jelly poured or smeared directly onto other foods, where it can retain its character and complement whatever it accompanies. Blue cheese is a particularly good mate.

The syrup will have a viscosity somewhere between simple

syrup and maple syrup, depending on how long you choose to boil it. It can be made by boiling fresh medlars, or by reboiling already boiled medlars that were used for medlar jelly. The flavor will be just as strong.

Yield: 1 pint

1 kilogram (2.2 pounds) well-bletted medlars, or previously boiled medlars

6 cups water (or enough water to cover)

Sugar

½ vanilla bean, slit open along one side (optional)

Cut each medlar in half, and place them—skin, seeds, and all—in a preserving pan. Cover with the water, then boil gently, with the lid slightly ajar, for 1 hour. Strain the pulp through a damp jelly bag, and let the juice drip for 1 hour into a bowl. Discard the medlar pulp (or reboil it for more syrup).

Measure the juice, then measure an equal amount of sugar. Combine the sugar and juice in a preserving pan, and cook over medium heat, stirring until the sugar has dissolved. Add the vanilla bean (if using), then increase the heat to high and boil for 5 minutes for simple syrup, 7 to 10 minutes for a thicker syrup—or until the liquid begins to drip viscously from the spoon when you hold it up for inspection.

Remove the syrup from the heat, and skim, cool, and bottle it, removing the vanilla bean if you wish. Store in the refrigerator, and use within a month.

N: Norton Grape

Vitis aestivalis 'Norton'
Vitaceae (grape) family
Also known as Cynthiana

I was going to write about Pinot Noir and call it "Noir, Pinot," so I could make it my "N" chapter—a little alphabet gymnastics maybe the reader wouldn't mind. Noir, Pinot, would be an excuse to write broadly about terroir and Portland, Oregon, how growing up in a certain place can make a person special like it makes wine grapes special, that you can't just move somewhere and grow different—or you can, happens all the time, a human thing as much as a plant thing. But the metaphor is obvious, and "terroir" is a problem. One of those words that sound ridiculous to those of us who aren't in the wine club. Like "mouthfeel."

Writing about wine in general is a problem. Same with mushrooms, two subjects whose fields of study are crowded with experts. I've never been one to enjoy shouting in a crowded room. I'd rather go to some off-brand place, the shitty bar next to the popular bar, and suck all the air out of that instead. Maybe because I'm from Vancouver—not the Vancouver you're thinking of, but the off-brand city adjacent to Portland that doesn't produce wine or hipsters or nationally admired restaurants, which makes some of us a little defensive. Pinot Noir is good wine, right? Sure, yes, of course it is.

Norton wine can be good wine, too, but when I tell our friend Kelly, a Certified Specialist of Wine, that I have three bottles of Norton I want his opinion on, he makes a face.

Norton, or *Vitis aestivalis* 'Norton', is native to North America. It's also Missouri's official state grape. If you don't already know this, you should: wine grapes native to the United States are not much loved by People Who Know Wine. Practically all of the wine most of us drink is from vines of European origin—*Vitis vinifera*—that are delicate and finicky, and often can't survive the extreme climates of many corners of the United States. California's okay. Oregon and Washington are okay. Texas, Minnesota, and Virginia, less so. Even in ideal growing regions like the Willamette Valley, *Vitis vinifera* is commonly grown on *Vitis aestivalis* and other *Vitis* rootstocks to help it thrive.

American grapes tend to be "foxy," a term that means musky and coarse, and is often related to a grapey flavor, like juice (I harvested this information from a Google search; Kelly says that it is broadly true but not really true, that "foxy" and "grapey" are separate characteristics).

Norton grapes are less foxy than Catawba or Scuppernong, and Norton vines grow well where *Vitis vinifera* doesn't, because they are disease-resistant—a live-or-die trait in humid climates like Missouri's. Most important, Nortons are slow to de-acclimatize. If we were to anthropomorphize the Norton, we'd say it's insensitive or mistrustful or supersmart, depending on how open-minded we were to non-European grapevines. Nortons aren't fooled by the first sixty-five-degree day that comes along. In a place with cycling winter temperatures like Missouri—warm one day, freezing the next—a plant needs to know when the weather's bluffing. Nortons appear to know.

Norton wines have a reputation for being sweet (code in more sophisticated wine circles for "bad"). In fact, Norton wines are usually dry, but the Missouri wineries that produce them, like Les Bourgeois Vineyard in Rocheport, have a reputation for sweet wines, such as Riverboat Red, a Concord grape blend. This popular sweet wine pays the bills, while Nortons remain a specialty purchase for connoisseurs who know better than to judge a winery by its bestsellers. That's one of the reasons Norton wine fights an uphill battle for industry respect, but that's not why Kelly sneered. He thought I was talking about the Argentinian winery named Norton, not the American varietal. Once he has tried my Nortons—his first, one of only three or four times he's tasted *Vitis aestivalis*–based wines, "a treat," he says—he tells me they have more in common with old-style European wines like Cabernet Franc than with the fruit-forward Cabernets and Syrahs many Americans know better. These European wines have a lower alcohol content (12 percent instead of 13 percent or more), with earthy notes and astringent personalities, he says. You can draw a relationship between the old-style, respected European wine and New World Norton wine. I'll buy that, I say, and take another drink.

Norton grapes don't root easily, but when they do, they grow voraciously, with huge leafy canopies that must be pared back. The grape itself is quite small, with a high skin-to-juice ratio (another attribute similar to high-quality European wines, Kelly says). Norton grapes also have a naturally high pH *and* malic-acid content, which is part of that grapey taste. Normally, the winemaker could deacidify grape juice by adding potassium carbonate. Norton wine-makers can't use that trick because it would only increase Norton's already too-high pH. Instead, they might put Norton juice through malolactic fermentation, a technique often used with red wine to transform the green-apple flavors of malic acid into the creamier flavors of lactic acid.

Hank Johnson of Chaumette Winery, in Sainte Gen-evieve, Missouri, solves this problem a different way, by using a Smart-Dyson Ballerina trellis system. This growing method increases sunlight exposure to the grapes, causing them to respire more malic acid. He explains this to me over prawns and wine in his tasting room, drawing a diagram in my notebook that looks like a muppet, vines flowing up and out and everywhere like wild hair, with a solid, skinny body (the vine) down below. He's poured me a glass of the Chaumette Reserve, which smells like my favorite Spanish wine, like dried fruit and fresh honey. It tastes good. Do I have to say what it tastes like? Like dusk on a European ter-race? Like driving your convertible through a construction zone? Like cherries and mushrooms and leather and nails?

Hank wishes there was a way to educate wine judges about how different the Norton is from California reds without resorting to professional PR, an idea that makes him look visibly grossed out when I suggest it. The wine can speak for itself, he hopes, if you listen.

When I ask Hank to teach me how to taste a Norton, he tells me that if his son were here he'd have all these words to describe the taste, more words than I'd believe. As his son said them, I'd be able to pick out all the nuances. The words would make each subtlety obvious. Hank doesn't have those words. Instead, he teaches me how to taste wine the way the pros do it. "Take a sip of the wine, tilt your head down, and suck in air, like this." He makes a snorkeling noise.

I try it. Wine rushes to the corners of my mouth and tries to escape. It rushes up my nasal passages, too, burning.

"Each sip should take thirty seconds," Hank says, which makes me wonder, with my mouth full, how we're going to talk about the wine.

It's weird, sitting with a liquid as if it were a lozenge.

I choke and sputter and attempt to make no noise doing so. Hank thinks it's funny, how bad I am at this. Confirmation bias, I think. That's one way wine words work. I say "dried fruit" and you taste raisins. I say "tobacco" and you taste pipe smoke. I say "aluminum" and you taste Diet Pepsi.

Sam and I spend that night in Sainte Genevieve, an old French settlement on the Mississippi, near Hank's winery. We know the river is somewhere but can't find it until, tracing the edge of town, we drive over a dike and find a burned-out cluster of buildings with graffiti so racist, and decay so complete, it is hard to imagine why the town would allow this building and graffiti to stand. I am inspecting the husk of what used to be a garage when a black SUV drives up. My instinct is: *Run*. So is the instinct of the teenage girls in the car. One of them springs out the passenger door, sprints to a doorway scrawled with BRING BACK SLAVERY, slaps the jamb with her palm, then runs back to the car and drives away. Just beyond a thicket of oak trees, the Mississippi River flows with barges and tugboats and birds, not the river of steamboat casinos and million-dollar views, but the river of sewage and dead bodies and burned-out buildings, where girls dare each other to touch the haunted house and arson's okay and the hate that overflows our online channels can burn in red spray paint on cinder block and wood.

Ah, we think. That could explain why, in the adorable wine-town, we only saw white people. That old rot at the center of the Union, that oxidized smell at the bottom of the glass, the bird-watchers in the parking lot of the burned-down bar that's outside the town limits—"not Sainte Genevieve's jurisdiction," as an archived issue of the local paper reports—so there's nothing town officials can do. With what words should we judge this place, we tourists who came to town merely for something interesting to drink?

Most reports say that the Norton grape came to Missouri from Virginia; that a couple years before the oenophile and enslaver Thomas Jefferson died, a Dr. Norton of Richmond accidentally hybridized *Vitis vinifera* with *Vitis aestivalis*, creating what we now call the Norton. For years, a Dr. F. A. Lemosy took the credit and marketed the vine under his own name, claiming he found the vine growing wild in a Virginia field a few years before Dr. Norton did. The Norton grape's provenance was further called into question when genetic testing performed in the 1990s determined that Norton and another native American varietal, Cynthiana, are the same grape.

Hank thinks that the Dr. Norton/Dr. Lemosy story is all hype, that, whatever these men made or found in Virginia, it's probably not the grape Hank's growing in Missouri dirt now. He prefers the theory that Norton evolved in Missouri and northern Arkansas, adapting over time to become the disease-resistant, hardy vine he now grows. He can't prove it, and as of this writing, a research team from Missouri State University reconfirmed in 2017 that the two cultivars are genetically identical. But that study did not settle the question of where Norton/Cynthiana comes from. For now, the researchers write, "the precise origin of the two cultivars can only be hypothesized."

But Hank knows what he sees. A grape that thrives in Missouri weather, Missouri soil; a grape that flourishes in this land of elegant wineries, tourist villages, and racist ruins. What is certain, regardless of name or origin, science or hype: this grape is American. This Norton is our grape.

Norton Gelée

Of all the fruits I've written recipes for so far, Norton grapes present a particular challenge: few readers will be able to source Norton grapes. I can barely source Norton grapes! I *could* write a jelly recipe for whatever wine grapes grow in your area (if wine grapes grow in your area), but this doesn't seem like a thematically cogent way to conclude an essay about a particular terroir. Anyone over twenty-one can order a bottle of Norton wine online easily enough (be careful you don't actually purchase a bottle of Norton Winery wine), but that presents a different problem: Why go to all that trouble to buy a fairly rare bottle of wine only to adulterate it in a recipe? Isn't the best thing to do with wine to drink it? Who needs a recipe for that? But my job here is to offer a recipe—one, I hope, that doesn't make the vintners cringe when they see how I messed with their product.

Norton jelly did not work. The high heat required to make a gel produced a preserve that smelled like wine but tasted like sugar. A gelée was much better. Gelée is *not* Jell-O. Here, gelatin stabilizes the wine without marring flavor, just barely turning it into a solid. It remains soft, spreadable, shuddery, the consistency of chilled bone broth or aspic. I had to add a little honey—without it, Norton gelée isn't palatable, or at least isn't palatable to my sweet-soaked American palate. In your gelée, you can forgo the honey or add it to taste.

Serve this gelée with Camembert, Curado, Roquefort, or some other delicious stinky cheese. Refrigerate leftovers.

This recipe makes one small ramekin of gelée, leaving four glasses of wine in the bottle to enjoy as its makers intended. For anyone used to big, obvious California and Washington reds (like me), Norton wine will be a new experience of more subtle flavors, best enjoyed with tart, herby food, like spiced sausages, roast poultry, or Indian cuisine.

Yield: 1 small ramekin of gelée

1 sheet PerfectaGel Gold leaf gelatin (do not use
    powdered American-style gelatin; it muddles the
    wine's flavor and produces a rubbery texture)

½ cup Norton wine

1 tablespoon honey

Neutral cooking oil, such as canola or grapeseed

"Bloom" the leaf gelatin by cutting it into 1-inch strips and soaking them in a bowl of water for 10 minutes. While you're waiting, combine the wine and honey in a small saucepan and heat them over medium-low heat, stirring, until the honey dissolves, then immediately remove the mixture from the heat so you don't alter the wine too much. When the gelatin has bloomed, pick it up (it will still be a little solid, and you may need to scrape it from the bottom of the bowl—I use my fingers), discard the extra water, and put the gelatin in the wine-and-honey mixture. Heat over medium-low heat, never boiling, and stir until the gelatin completely dissolves. Remove from the heat, and pour into a lightly oiled ramekin.

Refrigerate until firm, about 3 hours.

To unmold, immerse the ramekin to the rim in very warm water for 10 seconds, then remove it from the water, giving it a shake as you hold it upright, and upend it onto a chilled plate. The plate must be very cold; otherwise, the gelatin will melt. If this takes you to the limit of your culinary skills (as it does mine), just serve the gelée in the ramekin. Refrigerate leftovers. Best within 3 days.

Red Wine Vinegar

The root of "vinegar" comes from the Old French word for "sour wine," which itself comes from the Latin for the same term, *vinum* (wine) *acer* (sour). With the help of naturally occurring aerobic bacteria called *Acetobacter*, any wine can become vinegar.

Start with an abandoned or open-too-long bottle. Pour the wine into a nonreactive, wide-mouthed vessel (a bowl or crock, not a bottle), covering that vessel with a cloth to keep flies out but let aerobic bacteria in. Let the crock sit in a dark, well-ventilated spot. Increasing the surface area of wine exposed to air is important, because *Acetobacter* need oxygen to thrive. The number of weeks the wine needs to turn into vinegar will vary by season and by the amount of alcohol and sugar in the wine. In summer, vinegar making will go faster. In winter, slower. Higher-alcohol, low-sugar wines may need extra help. If, after a couple weeks, there's no activity (but maybe some mold!) in your red wine, and it still tastes winey, not vinegary, add a shot of raw apple-cider vinegar—any brand that still has active cultures. You can kick-start the process by adding raw apple-cider vinegar at the beginning.

If something starts to form on the surface of the wine, a film or a kombucha-like disc, that's the vinegar mother. It is harmless and edible, and can be used to start the next batch of vinegar. Strain it out, or eat it with the vinegar. If blobs form at the bottom of the vessel, these are by-products of the mother and can be strained out, left in, or used to start the next batch of vinegar. If mold forms, pluck it off.

Once a week, taste the vinegar. When it tastes sour (like the vinegar it now is), transfer it to bottles and seal. At this point, the vinegar must be protected from oxygen so it does not overferment. After the *Acetobacter* have converted all the alcohol into acetic acid, they will consume the acetic acid, exuding water and carbon dioxide this time, turning the wine they had previously turned into vinegar into a weird, murky water instead. Be cautious when opening a bottle of homemade vinegar. If it over-ferments, carbon dioxide will build up and the contents of the bottle could explode.

O: Osage Orange

Maclura pomifera
Moraceae (mulberry) family
Also known as hedge ball, hedge apple, horse apple,
bois d'arc, bodark, monkey ball

The first time I see them they're in a silver-wire shopping cart at the entrance to a Thriftway in East Des Moines, piled under the fluorescent lights like false starts in a wastebasket. The sign calls them "hedge apples" and invites me to buy them for 75 cents a pound. It isn't clear why I would, but I do. They look tough. Rumpled. Possibly inedible. But they *are* fruit.

"Trash fruit," Aunt Gail calls them ("Only you would spend money on trash fruit," she says), because Osage oranges—as they are also called—are indeed inedible, un-juice-able, and not useful as a home remedy, though it's easy enough to find someone to assure me they repel spiders.

Can we jam them? Ferment them? Boil their rinds or roast their fruit? The internet and Aunt Gail say we cannot. Gail suggests we throw them in the trash, where they belong. The internet suggests we use them as a decoration. Get a large glass bowl, pile the hedge apples high, and set them at the center of the table until they rot.

Why bother with inedible fruit? As with all the subjects I've chosen, I'm attracted to Osage oranges because they delight me and I want to know why. Biophilial feelings, E. O. Wilson would probably call them; "life's recognition of itself," as M. M. Mahood put it in *The Poet as Botanist*. But so what if I think Osage oranges are cool? Delight isn't a story. Appetite isn't a story. Anxiety isn't a story, either. If I were writing an encyclopedia, the facts would be the story. But that's not what I'm writing.

Start with its name. "Osage" comes from a French mispronunciation of Wah-Zha-Zhi, the Sioux tribe who prized Osage orange wood for their bows. "Orange" comes from the color of the wood, or maybe the bumpy rind of its fruit (though not from the rind's color, which is green), or maybe from how the fruit smells orangey when it rots. "Hedge" is what my friends Terry and Nina call it, though it's a tree. Hedge trees line the north and west corners of their property outside Columbia, Missouri, and provide shade for their bulls. Every now and then, Terry cuts them back, but they're easier to leave alone—especially in spring, when their sap runs so thick it destroys chain saws. It's a hardwood, rot-resistant, with a twisted grain that defies the ax. "Young farmers who want to build a fence to last their lifetimes might use hedge," Terry says. "But they'd have to dig giant postholes to accommodate the way the wood curves." When Nina wanted a path through the section of farm she was converting to a garden, Terry used discs of Osage orange wood. "It's such hard wood, *so* hard," Nina says. "Two hundred years from now, people will be throwing those Osage orange chips around like Frisbees." Terry loads the trunk of our rental car with fresh-split logs so I can burn them at my next destination. Their orange color glows the way natural red hair sometimes glows.

When separated from the tree, the fruit of the Osage orange is also called hedge apple or hedge ball. The fruit does not offer itself easily, like an orange wedge; its design is not "intuitive," as we might say of an Apple laptop's interface. The rind is tough and does not peel. One can extract the seeds from the fruit, but would need to wear gloves to protect against the latexlike juice that oozes from the pulp around them. One could roast the seeds, but they would taste only as good as any other, easier-to-gather squash seed. Nina can't remember a time when she didn't know about hedge apples, or a time when she interacted with them on purpose. They were always there, on the edge of her Kansas childhood, "these mysterious, slightly menacing things," she says. "I remember as a child trying to avoid them. I had this impression that they'd make me sick." She's lived with Terry on their farm for thirty years, but the first time she's visited the trees is the day I ask to see them.

Hedge apples show up every now and then on *Martha Stewart* as a supercute holiday decorating idea. Martha recommends stacking them with quince and figs on glass pedestals with dried leaves. "The effect is *exactly* what one wants at this time of year," she says.

Martha Stewart is right—hedge apples *do* look cute. They're bright green, patterned like brains, as big as softballs, and scentless until they rot. Some people like their citrusy rot-smell. "Asshole snowballs," my friend Kate calls them. They've been recommended by farmers as fodder, but they've also been known by farmers to choke livestock. A hedge-apple-derived beauty serum (developed by the chemist son of an Iowa farmer who never could find anything to do with his hedge balls) now retails for $100 an ounce. It taps into a homeopathic, like-cures-like logic that's particularly potent when applied to anxieties about aging and death: If this tree won't rot, neither will you. Apply to the face with your fingertips, using a light, circular motion.

The thing about hedge, Terry shows me, is that even when a branch grows too heavy and twists off in a storm, even when a tree topples over, new branches will grow from that base. He points to a trunk that runs along the ground like a monster root. Five new branches are growing from it toward a sunbreak in the canopy, so straight they look like individual saplings. And along each "sapling," nasty-looking thorns that an herbalist I interviewed the previous day told me will grow into new branches. A stirring image, but untrue.

The last creatures to understand Osage oranges might have been the megafauna that coevolved with them. One hypothesis speculates that when these giant sloths and beavers died out they took the tree's original method of seed dispersal with them, limiting its range to a boot-shaped stretch of East Texas and southeastern Oklahoma. The spread of Osage orange in the Anthropocene began when it was planted by settlers to fence government-granted land before the advent of barbed wire. It spread farther when the Works Progress Administration encouraged planting it as a shelterbelt. Hedge apples were a by-product of good fences, of marking territory. Not a crop anyone wanted.

We've driven on to Kate and Brian's house in Columbia, where they let me burn Terry and Nina's Osage orange wood in their backyard. Some of it has been cured—we can tell by its faded orange—and some is oozing sap where Terry cut it this morning. Brian and Kate and their daughter, Lulie, and I collect rocks from the landscaping and make a circle in the clover while Sam takes our car to get a blown tire fixed. Brian starts the fire with kindling Terry packed for me and feeds it slightly larger sticks, piece by piece, until the fire is big enough to handle a log. While we wait for the blaze to grow, Lulie collects pokeberries and smashes them with her mortar and pestle. She returns with pink pokeberry ink on the dress she changed into after getting pokeberry ink all over the first, and all over the sleeve of her mother's sweater, which her mother isn't bothering to change out of. Her father is building a fire for me on their back lawn because I am not good at building fires—or at being outdoors—and I want to know if what the internet says is true: that Osage orange fires burn so hot they let off fireworks.

This possibility—like the claim that hedge apples repel spiders—appears here and there online, and seems to me like the kind of folk wisdom that gets passed down by grandparents and recycled into decorative truths by life-style magazines.

Brian feeds cured logs to the fire while we watch.

The Osage orange fire *does* spark—like a fire. Shards of wood peel away in the heat and combust, sparking in the air as one would pretty much expect.

We watch Osage orange burn until Sam comes back with a new tire. Then we all pile into Brian's Toyota to go trespassing.

When I think of exploring, I think "nature trail" or "city sidewalk." Brian thinks "abandoned train tunnel." He's never happier than he was as a boy, Kate says, when he could wander his family's Missouri land with a gun strapped to his shoulder and a stick to ward off stray dogs.

Lulie's rules of trespassing, as learned from her father: Don't steal. Don't damage anything. Be respectful. Be quick. Don't get caught. If you do get caught, leave. When she strikes out on her own and gets stuck, Brian dives into the brush and backs out of it, arms wide, branches and vines spreading over his back before snapping, Lulie marching through the clearing his body has made for her, yelling, "Bushwack! Bushwack!"

Brian leads us to a train tunnel hewn through a hill. The tracks have been removed, leaving a gravel bed pocked with puddles. Not that we know about the puddles yet. From here, we can't see what might be inside the tunnel besides a darkness so dark it looks like it has mass and texture, and a far-off light that looks like a doorway. Together, we are going to walk through that door. "Where'd you hear about this place?" I ask Brian. "The internet," he says.

It is not that I hate the outdoors or the dark. It's just that I know that if I step off the trail I will be killed by something that deserves a good meal.

I feel this on a cellular level, in my bones, at the base of my brain. I must have been a barn mouse in a past life. Or the horse that bolts for the barn.

We walk into the dark, decommissioned train tunnel. The ground seems to rise to meet my feet, my vision so panicked by its lack of perspective that it invents a new horizon. Kate forgot to exchange her sandals for hiking boots, and when I hear her splash into a puddle I feel a shiver of empathy for something she might not even care about, that the tunnel is *touching* her. And ruining her shoes.

If there is a monster in this tunnel, it is my desire to turn around. If there is a monster, it is the thought that after we walk through this tunnel we'll have to turn around and walk back.

Kate does not appear to care if the tunnel is ruining her shoes.

About halfway through, Brian veers into a maintenance alcove that feels more like the plumbing of my psyche than a physical space. I hold Sam's hand. Lulie is afraid, but she does not reach for her mother or father.

///////////////////////

We keep walking. Water splashes our ankles and drips from the ceiling, and there is no warning when the water hits or the foot stumbles, just walking. When I look to the daylight far ahead of us, its afterimage blots out the darkness closer in.

What is the point of this walk? To see what's in the tunnel—or on the other side? To break a rule that doesn't matter? To hike through the dark together, with friends, for the hell of it? When my aunt was dying in Iowa, my mother flew from Washington State to be with her. She took my brother, while my father and I stayed home. What I remember of that time was a moment in the morning when I was almost ready for school and needed my father to fix my hair. He took the metal clip I gave him, the wide tooth of it open to catch my bangs into the sideswipe that would keep them out of my eyes. As he drew it from my temple to my ear, I could feel his right hand shake, uncertain on the clip, not holding my head still with his left hand as my mother would have done, trying to capture my hair as I had asked him to, doing all right, but not well. *He doesn't know!* I thought.

I can hear Lulie ahead of me. Is she still afraid?

I offer her my hand.

She says no. "No, thank you," Lulie says. "I have to learn."

Spider Balls

Osage orange's most enduring use is also its most enduring myth. Folk remedies, grandmas, and that guy who grew up in the Midwest and was delighted to hear someone mention hedge apples will all likely say hedge apples can be piled around foundations and in basements to repel spiders. Research at Iowa State University on these repellent properties found that, yes, Osage oranges do contain a chemical that discourages cockroaches, flies, and mosquitoes, but only if cut in slices and left in small, airless rooms. Intact hedge balls arranged anywhere there's air circulation don't release enough chemical to have much effect. When trying this folk remedy, place these "spider balls" on plates or some other surface that will catch strange goos and stains as the fruit rots, and keep an eye out to make sure they aren't actually *attracting* bugs. Should you decide to risk a blade on cutting into one of the balls, wear gloves. The fruit's juice irritates skin.

Osage Centerpiece

Those who live in the Midwestern United States might think of hedge balls as trash fruit, but outside that region it's almost impossible to get a supply without paying a ridiculous amount for the privilege. So, for the lucky Osage-orange-rich people out there: When they're ripe in September and October, pick some or gather them from where they've fallen to the ground. Take them home. Put them in a bowl. Set the bowl at the center of the table. Take a photo for the rest of us.

P: Pomegranate

Punica granatum
Punicaceae (pomegranate) family

When Persephone returns to her mother, the underworld is still on her. In one version of this mythic reunion, Yannis Ritsos writes:

> *I heard you all calling my name;*
> *and my name was strange; and my friends were strange;*
> *strange the upper light with the square, pure white*
>   *houses,*
> *the fleshy, multicolored fruits, pretentious and*
>   *insolent . . .*

Persephone has seen the dead, married their king, eaten three or four or seven seeds of his pomegranate. Her mother, the harvest goddess Demeter—having been flattened by grief, having refused to let new crops grow until her daughter returns, having starved mortals until the gods fear no one will survive to leave offerings, having, in another version of the myth, convinced Zeus to make Hades give Persephone back—welcomes home a changed girl, wizened and spooky, uneasy in her mother's empire of green. A married girl who hears and speaks of a world Demeter can't understand. "The voice is paler than the lips it leaves," says Demeter in Edith Wharton's retelling, her joy fading to confusion.

Pomegranates are unusual fruits, "no more than a closet of juicy seeds," as Jane Grigson describes them. Poets have been known to compare those seeds to jewels. Cracking open a pomegranate does feel a bit like lifting the lid of a jewelry box, in expectation if not sensation—unless one tears open a jewelry box in a defensive posture, anticipating a spray of red. Within the split rind, an ornate pattern, edible and glistening.

According to Jewish lore, the pomegranate contains 613 seeds, one for each mitzvah. For millennia across Europe, Persia, and Asia, in Buddhist, Islamic, Judaic, and Christian traditions, pomegranates have been invoked as a symbol of fertility and sometimes smashed in bridal chambers to encourage the birth of many children. In *The Unicorn in Captivity*, a medieval European tapestry one can inspect before touring the quince grove at the Met Cloisters in Manhattan, a unicorn sits within a low-fenced pasture beneath a pomegranate tree. He looks content in captivity, a symbol of fertility and marriage and the fertility of a soul's marriage with Christ. The unicorn appears to be bleeding from wounds of the hunt that chained him to this tree. On closer inspection, the wounds don't bleed—they weep seeds. The blood is pomegranate juice.

Pomegranate seeds are incisor-shaped—fat at one end, where a blood blush pools, narrowing at the translucent tip, where the seed might, were it an actual tooth, root in the jaw. If we believe the Doctrine of Signatures—the idea that God has written a language in plants that we can read to identify our medicines—this shape means pomegranates can relieve oral maladies. "A strong infusion cures ulcers in the mouth and throat, and fastens teeth," wrote Culpeper.

It's strange to us now, this sort of anthropomorphizing that dismembers plants into humanlike parts instead of giving them humanlike personalities—tooth-shaped tooth fixers, not *The Giving Tree*. The Doctrine of Signatures was part of the worldview by which early doctors, herbalists, and apothecaries transformed an organism into a specific medicinal resource, an alchemy we present-day capitalists surely understand. "Each plant was a terrestrial star," Agnes Arber describes in her 1912 history of herbals, "and each star was a spiritualized plant." Modern marketing bypasses the Doctrine of Signatures and mines Greek myth instead, selling pomegranate juice as an elixir of youth, with antioxidant promises that fall just shy of raising the dead.

In my glass right now, as I write this: iced pomegranate juice and the black-winged corpse of a fruit fly. The juice is sweet, acidic, and tannic, wicking the moisture from my mouth in a pleasant way, a quenched feeling that also makes me want another drink. Tannins in pomegranate juice, as with good wine, balance acid and sugar and add a sense of chewiness and substance, like I'm eating something from the earth. Juicing a pomegranate can be as easy as pressing the fruit between your palm and a countertop, crushing it gently as you roll, then cutting off the top and inserting a straw. Eating pomegranate seeds requires a bit more work. Start by scoring the peel, then pull the fruit into quarters, revealing garnet-colored seeds. The word "garnet" comes from "pomegranate," as does "grenade," so named for the way a shrapnel-scattering grenade imitates the seed-scattering explosion of a smashed pomegranate.

In the myth, or a version of it—all versions of it— Demeter mourns her daughter's disappearance by letting the crops die. She abandons her duties and walks among mortals disguised as the sort of old woman who might look after the children at court. Nothing will grow until her daughter returns. And even after Persephone comes home, she has eaten the food of the dead and must go back to Hades for a fourth or a third or half the year, provoking another winter. This cycle of death and rebirth makes Demeter and Persephone empathetic to mortals as no other gods are. "In their grief and at the hour of death," Edith Hamilton writes in her 1940 anthology of Greek myth, "men could turn for compassion to the goddess who sorrowed and the goddess who died."

Pomegranates represent fertility, but also a pause in fertility—in myth and in life. In ancient Greece, Dioscorides recommended pomegranate seeds and rind as birth control. "Medical writings indicate that pomegranate was administered as a suppository," John M. Riddle writes in *Eve's Herbs*—not orally, as the myth might lead us to conclude. He reports that in 1933, date palms were the subject of the first experiment that found estrogenic compounds in plants—the first confirmation that herbal-birth-control lore had a biological and scientifically measurable basis (though the experiment's results were not duplicated and confirmed by peers until 1966). Subsequent experiments in the 1970s and '80s on the contraceptive powers of plants found that female rats fed pomegranates and paired with male rats who were not fed pomegranates experienced a 72 percent drop in fertility. In guinea pigs, the drop was 100 percent. The seeds, roots, and whole plant had no effect; the estrogenic compound was in the fruit—specifically, the skin around the seed and the rind. After forty days off the pomegranate diet, the rodents' fertility returned.

Spring arrives with death still on her.

> *It hurts to be born.*
>
> (EMILY KENDAL FREY, from *Sorrow Arrow*)

Frühjahrsmüdigkeit: the German word for an emotional state that sometimes translates as "spring fever" but is better translated as "spring fatigue," a mood disturbance that people like me say occurs when spring does that thing it does with Neruda's cherry trees. *Früh* (early) *jahrs* (year) *müdig* (tired) *keit* (ness). "Quiero hacer contigo / lo que la primavera hace con los cerezos." A split and froth of blossoms that makes me feel like breaking open, too.

///////////////////////

After four years of standing on a low bridge to worship the Spokane River's spring run, I understand that I crave its green crash not because I want to jump but because spring is violent, resurrection is violent, being born is violent. Spring makes me want to *be* the river, a stream of particles flowing to the sea with no identity or personality, belonging to everyone and no one, being everything, just matter mixed with matter, cutting a bed from basalt while flushing heavy metals to the sea. People do jump, they do die, and it is awful. No one survives the biggest drop, the lower falls in the middle of the city that even salmon couldn't climb back when salmon ran like a river within this river, before the completion of the Grand Coulee Dam, which, in 1939, after tens of thousands of hundred thousands of millions of years swimming home, blocked them from their downtown spawning grounds. If you live in Spokane today, you do not know a river of fish, but you do know the peculiar crawl of a particular kind of traffic jam, that standstill that means someone jumped.

The difference between craving death and craving transcendence—Persephone doesn't always know it, either. It takes her a little time to come back to life.

By fall, she misses her husband.

In some retellings, Zeus instructs Persephone not to eat while she's in the underworld. By the time Hermes retrieves her, she's starving. Hades offers her his pomegranate.

In Rachel Zucker's *Eating in the Underworld*, Persephone leaves Demeter by choice,

> *Away from where the body*
> *of my mother is everywhere,*

a journey that mimics the mature (but still painful) detachment of daughter from mother, who, because her mother is everywhere, must go to Hades—both a god and a place—to get free.

> *Only a mother could manufacture such a story:*
> the earth opened and pulled [me] down.

In this version of the myth, pomegranates represent the persistence of life, but they also create the marriage bind that demotes the primacy of mother-daughter relationships and interrupts fertility. Winter, in this story, is watching your daughter grow up into someone you can't understand. It's escaping your mother so you can know yourself without the crush of her fertility and love.

Winter is also rest. Demeter mourns and refuses to work. With fertility paused, a farmer can rest from the frenzy of planting, tending, harvesting, selling, preserving, and storing before planting again.

> *Remember, when you see me,*

Persephone says,

> *I am inside who I was.*

The earth where we raise our crops is the dirt where we bury our bodies. Pomegranates represent this same contradiction, this complete cycle: life and death and life again, returning new, returning transformed.

Pomegranate Molasses

Technically, pomegranates are berries, with fruit forming as juicy skins around each small seed—also called arils—separated in sections by a pithy white albedo and protected by a firm rind. If you're the sort who enjoys fussing in the kitchen, you'll find the labor of prepping pomegranates pleasant.

The quickest and messiest route to the arils is a cleaver—just cut the thing in quarters and pick out the seeds. For a tidier, more precise method, run cold water into a large bowl. Score a shallow circle around the calyx (the part that looks like a nipple), then get a firm grip on the calyx and pull the bottom away (off the tree, this bit of anatomy will look like the top). Score the peel where you can see the white pith separating the arils into sections, then break the pomegranate apart underwater. Using your fingers, loosen the arils from their padding of albedo. This will feel very much like unpacking a fruit rather than peeling it. The arils will sink; the pith and peel will float. Strain out and discard the inedibles, then drain the water from the arils. Then eat them.

Or make pomegranate molasses.

Yield: 1 cup

Arils of 8 pomegranates (about 9 cups), or 4 cups prepared pomegranate juice

½ cup sugar (more or less, depending on how sweet you'd like your molasses)

2 or 3 tablespoons lemon juice (more or less, depending on how tart you'd like your molasses)

Whir the arils in a blender on a low setting until they give up their juice, but not so long that the blender starts to mash up the seeds. Strain through a sieve, smushing the last of the juice from the seeds with a muddler. Reserve the juice, discard the seeds.

In a wide, heavy-bottomed saucepan, combine the pomegranate juice, sugar, and lemon juice. Bring to a boil over medium-high heat, stirring to dissolve the sugar, then reduce the heat to maintain a simmer.

Simmer until the juice is thick like syrup and has reduced by more than half. Add more sugar or lemon to taste, if you want. The molasses is done when you can see the bottom of the pan as you stir the syrup. Expect the syrup to have about a quarter of the volume of the juice you started with.

Store in the refrigerator. Drizzle over salads, cheeses, meats, anything that could use a little tannic sweetness.

Semi-Deathless Pomegranate Mask

Pomegranates are celebrated by beauty writers for their antiaging, antioxidant properties. The only time reversal I can confirm from this pomegranate mask is that it gave me acne. It does give the face a pleasant cooling feeling. And it tastes good. Should this beauty treatment fail to enliven your complexion, you can eat it for breakfast.

    1 tablespoon pomegranate arils
    1 tablespoon full-fat plain yogurt
    1 teaspoon honey

Pound the ingredients together with a mortar and pestle. Using a facial brush (I used a clean pastry brush), apply to the face. Let dry for 15 to 20 minutes. Rinse off with warm water. Drink the leftovers.

Q: Quince

Cydonia oblonga
Rosaceae (rose) family

How do I describe this? One October day, a friend brought a sack of quince to work and told me it was mine if I wanted it. The fruit was yellow under a scrim of gray fuzz, voluptuous and firm, like pregnant pears. I reached in, pulled one out, held it to my face. I sniffed the stem ends and shoulders, then turned the quince over and smelled the divot where a brittle brown star echoed the fruit's former blossom.

I could say this quince smelled like roses and citrus and rich women's perfume, but that isn't quite true. I could call this fruit "the stranger," based on what John Gardner called one of two possible plots in all of fiction—"a stranger comes to town." ("Man goes on a journey" being the other.) Calling quince "the stranger" could be fitting for a tale about the fruit of rooted things written by a woman whose female forebears did not make journeys. Instead, they turned inward: my grandmother at her TV; my other grandmother in her garden. My grandmothers escaped fathers and siblings by falling in love. They married and moved away and labored in little suburban houses where they did not speak of God, but gathered His artifacts around them, crosses and figurines that claimed the chapel of the living room, the sanctuary of the lawn.

I inhaled this stranger, my first quince, until my nose lost track of honey and citrus, but still held a wisp of cool, clean peel, an idea of sweetness, a thin hit of rose. To get my pleasure back, I could have waited for my senses to calm down. Instead, I did as the fruit's generous shoulders seemed to suggest.

I bit one.

The quince was firmer than an apple. I felt, for a moment, like I was using my teeth as a knife. Then an astringent sour sensation wicked all the moisture from my mouth. I stood dumb, cotton-tongued, the quince loose in my hand. I'd expected it to taste the way it smelled.

"The rabbinical traditions of the Jews make it the most ancient of all our fruits, dating back to the Garden of Eden," writes William Witler Meech in *Quince Culture* (1888). "There, by its exquisite beauty and delightful fragrance," the quince tempts Eve "to commit her first disobedience," the bite of voluptuous, sour fruit that opens her eyes to good and evil and exiles her descendants from paradise. *Tappuach*, the Hebrew word for apple, may also refer to citron, apricot, or quince; many scholars argue for quince because it is the only fruit of these four that is native to Persia and Palestine. The Song of Solomon, then, does not comfort with apples.

Nor are quince ready for consumption the moment they are picked, the way apples are. A quince needs the heat of a sauté pan or the roasting of an oven to become suitable company for a meal. It is a cook's fruit. It is difficult.

As quince stews, its cream-white flesh turns deep rose, and its fragrance transforms from something heavenly to something earthbound but still delicious. Quince has a satisfying grainy texture when cooked, like a pear that has kept its composure. If you've seen quince before, it's likely been in an extreme version of this state, as membrillo, also called quince paste or fruit cheese, which can be made by cooking the fruit with sugar over low heat until it is thick and concentrated and rose-colored, cooling it in a terrine, then serving it with Manchego and bread.

Quince can take this solid shape because its flesh, seeds, and skin contain large quantities of pectin. Place quince seeds in a cup of water and walk away. When you return, the cup will be thick with gel.

Some internet sources link the decline of quince production to the invention of pectin extract for household jam-making, which happened in 1908, but as far as I can tell this is a rumor. Quince was losing friends before Pomona's Pectin came along. It was so unpopular twenty years before, in 1888, that William Witler Meech was moved to write the first comprehensive guide of quince before it fell into obscurity. By the time my father's mother was born, in 1922, quince wasn't a secret to be discovered or a mystery to be solved. It wasn't even a rumor. She'd never heard of it. It didn't exist.

"I shouldn't say this," my grandmother Loretta admits, "but I was not happy when I got pregnant three months after I had your uncle. I was sick every day of all of my pregnancies."

My grandmother has, in a way I am by now familiar with, separated her love for her children from her personal frustrations, so that they run on parallel tracks of what she endured and how she felt about it. Whom she was pregnant with—my father—is a detail we can skip. Any unhappiness she felt was not because of him.

Eight years after my dad was born, my grandparents adopted two more children, girls we do not talk about.

The day I found out about them, I was sitting on the floor of my grandparents' basement, cross-legged behind a bar whose cabinets were stuffed with boxes of photos. The sun filtered down from the top of the room, strong at the ceiling but weak at the floor. When I came upon a few snapshots of little blonde girls, I had to stand to see them clearly.

I didn't think much of them at first. They could have been the neighbors' kids.

Then I found a family portrait, shot with the soft light and neutral smiles of a photography studio. There was my father when he was my age at the time, eleven years old. There were my uncles, twelve and twenty-five. There was my grandfather. And there, on my grandmother's lap, two blonde babies.

That day, I studied the photo for a minute, then put it

away. I put all the photos away. Who were those girls? I did not ask my grandparents.

Who were those girls? I did not ask my father.

"Who were those girls?" I did ask my mother, weeks later.

"Your aunts," she said.

Pairidaeza—ancient Persian for "place surrounded by walls"—is the root of the word "paradise."

In the Middle Ages, Eden was imagined as an actual place at the edge of the map, and remembered—as it still is in Judeo-Christian and Muslim cultures—as a lost paradise where man's innocence gave him a direct line to God. Walled gardens in private homes and monasteries mediated between indoor space and the rural or urban wilderness outside, providing an in-between enclosure. In Europe, these gardens were planted with medicinal herbs, space-shaping hedges, and fruit-bearing trees. Within these walls, quince trees were common. They were one of the first trees to bloom. Their branches visually bridged the sky and earth. The pectin within their flesh firmed the household's runny jams.

A Christian version of these gardens was called *hortus conclusus*—"enclosed garden." Such enclosures symbolized the Virgin Mary's womb, her innocence as she prepared to receive the seed of God. This enclosed garden was an echo of paradise; constructing, maintaining, and sitting within one were all ways to channel prayer and thought and reach past human frailty. A quince tree cloistered there symbolized Christ, the tree of life.

A bite of quince opens Eve's eyes, so God throws her out

of paradise. When I was a young girl, cheering for Eve's bite felt like an open secret. Who among us wouldn't make the same choice? Who among us didn't think you'd be an idiot to abstain?

They were bad girls, my aunts. When I ask family about them—first as a child, then again and again over the years, whenever it seems we're in the mood to talk about tender subjects—this is the feminine shape the truth takes, an answer to my question that still doesn't tell me what I want to know: bad girls who left the family; bad girls, cast out.

Before I discovered my aunts, I'd thought we were less troubled than other families. Now I knew: It was possible to disappear from this family. To be loved and raised and erased.

The summer before I went to college, I finally asked my father about his sisters. They did not obey the rules of the house, he said. They left, he said. He'd been away at Iowa State then, about to move to the West Coast with my mother. He didn't know what had happened.

I was almost thirty when I asked my elder uncle directly what the girls had done. "They did not obey the rules," he said. "They were bad girls."

My grandmother has acknowledged their existence to me only once. My grandfather was sick at the time, maybe dying. We were talking about getting used to that idea. "I won't be happy about it when it happens," she said. I asked if they wanted to see their daughters before he maybe . . . possibly . . . We weren't quite saying "died." "No," she said, grimacing like I'd served her something sour. "I do not."

What could a teenage girl do that still looks monstrous

decades later? How could some girls go so bad that we still can't describe it?

Far past my teenage years, the dangerous age, I asked my father, "What do we have to do in this family to be kicked out?"

"Nothing," he said. "There's nothing you can do."

I remain protected, enclosed.

A good girl.

Eve told the serpent, "We may eat the fruit of the trees in the garden; it is only about the fruit of the tree in the middle of the garden that God said, 'You shall not eat it or even touch it, lest you die.'"

"You certainly will not die!" the serpent said.

The fruit was curvy and firm, bright yellow under a gray beard. It smelled like roses and citrus and rich women's perfume, like nothing she'd ever smelled before.

She pulled it from the tree and ran her thumb over the skin. The fuzz came off like dust. She held the fruit to her face. She took deep breaths at its stem and curves and bottom until its scent overwhelmed her, and wasn't enough.

Quince Jelly and Membrillo

During my initial research into quince, I encountered the notion that some quince need to be bletted—left on the branch in the frost, or left in the basement until they soften—in order to become edible. This quality made sense to me, considering that the astringency of raw quince is a lot like the astringency of medlar and persimmons, two fruits of almost-winter that definitely need a good bletting. I liked this "quince need to be bletted" idea so much, I let it tag after all versions of this chapter until this final version, when I realized that, after a near-decade of harvesting quince from maritime climates and more extreme inland climates, buying quince from specialty grocery stores, and scouring cookbooks for the best way to prepare and preserve them, I haven't found a single recipe that uses bletted quince, nor have my quince ever bletted in a way that makes them more edible, losing their astringency and getting sweeter, the way medlars do. My quince wrinkled, darkened, and rotted, like apples or pears. They took longer to rot than apples or pears, but, still, they rotted.

In the absence of strong oral traditions about food—and in general—I sometimes turn to food blogs and Wikipedia to find folk knowledge about fruits. As much as possible, I try to trace how this knowledge is intertwined with verifiable facts. This relationship can help me understand the cultural roles of these fruits, the way a supposed quality invites a metaphor that gives the fruit extra significance. I pulled this "fact" about bletting from Wikipedia, which as of this writing still notes, "Most varieties of quince are too hard, astringent and sour to eat raw unless 'bletted.'" Don't believe it.

There starts to be a sort of tide to quince jelly once it's close to being done, a shuddery pink primordial motion to it—but even then, I'm never quite sure if it's done. It has a snotty quality that other jellies don't have. Instead of accumulating in crinkles at the bottom of an upheld spoon, it lobs itself off in drops that stretch like spit.

The best way I can tell that the quince jelly is done is to wait until it seems like it's almost ready to burn, and quickly take it off high heat. Something deepens in the color and smell, and my fear ratchets up—that's the only way I know the right moment. This works in practice, but it's useless as an instruction in a recipe. Imagine it. "Step Five: boil hard until you feel a little sick to your stomach." As I tug my spoon through it, the jelly tugs back, then slams forward in the direction I was pulling right when I start pulling in the other direction. The final product shimmers and shivers like gelatin, but does not shear like gelatin. I have to push my finger firmly into the surface to scoop any up, yet the divot I make will slump into itself and disappear. It has viscosity but not stability—it's more like malleable candy, sticky and sweet. Maybe I've cooked it too long after all.

This time it doesn't matter, because most of my quince jelly won't be eaten as jelly: I'll use it as pectin stock for runnier jams. Melt 8 ounces over low heat, then slide the liquid jelly into a peach preserve and boil as usual until the jam sets. There's no reason to use this pectin instead of powdered pectin purchased from the grocery store, but I like the accidents of taste and texture that happen when I refuse the easier route.

Every part of quince is rich with pectin, the seeds especially so. You can tie the seeds in a muslin bag and add them to the boiling juice and sugar to help firm up your quince jelly, but it will be firm with or without them. Or save this trick for other, less pectin-rich jellies, like grape or medlar.

After you filter the quince juice through the jelly bag, you can refrigerate the juice and return to it up to 3 days later to continue the rest of the recipe.

Yield: 36–48 ounces

1 kilogram (about 6 cups) peeled and cored quince, reserving (optional) seeds

1½ kilograms (about 6 cups) water

Sugar

Juice of ½ lemon

½ vanilla bean, slit open along one side (optional; do not
substitute vanilla extract)

Cut the quince into 1-inch chunks, and place all the seeds (if
using) in a piece of muslin or cheesecloth tied with a string.
Combine the quince, quince seeds (if using), and water in a
heavy-bottomed pot, and boil gently with the lid partially on, not
stirring, for 45 minutes. Skim any scum from the top that appears.

Remove the pot from the heat, then filter its contents through a
moistened and wrung-out jelly bag into a medium bowl, discarding the bag of quince seeds. Let the pulp hang for 2 hours.
Do not squeeze the jelly bag to help the pectin run through;
this makes a cloudy jelly.

Prepare a deep canning pot with enough boiling water to cover
enough jars to hold 48 ounces (six 8-ounce jars, for example). I
add a little white vinegar, to mitigate the powdery white residue
my hard water leaves on my jars after I boil them. Sterilize the jars
by keeping them immersed in the boiling water for 10 minutes,
then set them on a clean towel to cool and dry. Boil the lids for
10 minutes, then place them on a clean, lint-free towel, seal-side
up, to dry. Rinse the bands and set them aside. Keep heating the
pot of water on low until the end of this recipe, when you'll use
this water bath to process the filled jars.

Reserve the quince pulp in the jelly bag for making membrillo,
and continue with the pectin stock in the medium bowl. You'll
have about 4 to 4½ cups stock—maybe more, maybe less.

However much pectin stock you get, pour it back into the heavy-
bottomed pot—an enameled Dutch oven will do well here—
bring it to a boil over high heat, and add the same volume of

sugar to it, plus the lemon juice. Stir to combine. I had 4½ cups pectin stock, so I added 4½ cups sugar. If you intend to eat this quince jelly as it is and not use it to thicken other jams, add the vanilla bean.

Boil hard, over high heat, without stirring, until you feel a little sick to your stomach, 10 to 12 minutes, skimming off scum as needed. The jelly will foam up, then boil madly within the foam; then the foam will start to shudder a bit and might darken slightly. That's a good time to test the set. The freezer method doesn't work for me with jelly—I just cover a wooden spoon in the jelly and set it aside on a plate for a minute (I take the jelly off the boil while I do this). If it wrinkles when I come back and push my finger through it, and if it gathers at the edge of the spoon before dripping snottily/viscously, the jelly is done.

Once you've achieved a set, remove the jelly from the heat and ladle it into the clean jars, leaving ½ inch of headspace. Wipe the rims clean, place the lids, and screw the bands on, fingertip-tight (tight, but not so tight you need strong hands to reopen them). Bring the water bath back to a boil and process the jars, keeping them immersed for 10 minutes. Remove the jars from the water, and set them aside to cool. Before storing them away, check their seals.

To make membrillo, run the pulp left over from making quince jelly through a food mill's medium disc. Take the quince pulp, measure it, and add the same amount of sugar to the pulp. In a preserving pan, cook the sweetened pulp over medium-high heat, stirring occasionally, until the mixture becomes so thick you can see the bottom of the pan when you trace your spoon through the pulp, and it piles up on itself when you drop some of it from the spoon back into the pan. It's hard to say how long this will take, since the timing depends on how much liquid was in your pulp. Keep an eye on it so it doesn't scorch.

Then, in a lipped baking sheet or casserole dish lined with parchment paper, spread the hot pulp evenly. I have the best results when I spread mine about 1 inch thick. Let the membrillo dry overnight. When the top feels dry and solid-ish, flip the paste over onto a cutting board that's been lined with parchment paper. Peel off the top (once the bottom) parchment paper, and let the bottom dry. Repeat this process, flipping the membrillo over and back, until the membrillo is firm enough to cut. Store the membrillo by wrapping it in parchment paper and keeping it in the refrigerator, or freeze it. It will keep in the fridge indefinitely but will become grainy and crystallized over time. Eat within a month for best results.

Bandoline

Columella, of first-century Rome, regarding quince: "The ripe fruit eaten raw is said to be good for spitting of blood; also for swollen spleen, dropsy, and difficult breathing." In addition to its medicinal uses, quince seeds were used cosmetically in a hair pomade, which pharmacists concocted into the nineteenth century by soaking pectin-rich quince seeds in water. The same recipe can be diluted with water, lemon, and honey to make a tea that soothes sore throats.

According to William Witler Meech: "Bandoline is made by covering the seeds with forty to fifty times their bulk of warm water, which soon produces a mucilage used by perfumers and hair dressers. Many ladies prepare it for themselves to keep their hair in place. It can be perfumed with any kind of odor. By the addition of a little alcohol it can be kept for a long time. It is this use of the seeds which causes the great demand with druggists."

For a modern adaptation, collect 20 grams seeds and immerse them in a bowl with 800 grams warm water. Leave this on the countertop and let it sit a couple hours, or up to overnight. When you return, you'll find the seeds sitting in a thick cloud of mucilage with a layer of water on top. Pour the contents of the bowl through a strainer, letting the water wash away, then quickly put another bowl beneath the strainer to collect the mucilage that will slowly drip out. To speed the process, stir the seeds and mucilage in the strainer until most of the goo has separated from the seeds. You probably won't be able to get all of it. Discard the seeds. Refrigerate the goo. Keeps for 3 weeks. Use as a gel to control and shape hair, or add to tea to soothe sore throats.

R: Rhubarb

Rheum rhaponticum (false rhubarb),
Rheum palmatum (Turkey rhubarb or China rhubarb), and
Rheum x hybridum (culinary rhubarb)
Polygonaceae (knotweed) family
Also known as Turkey rhubarb, Chinese rhubarb,
bastard rhubarb, garden rhubarb,
sweet round-leaved dock, pie plant

Rhubarb stalks are red or they are green or they are both. If they are green, they will never be red: they will not ripen because they are already ripe. They have achieved the quality "ripe" often measures—ready to eat. Harvest the stalks once you are satisfied with their size and do not wait for them to blush.

Do not eat the leaves. They are sour, like sorrel, and contain oxalic acid, also like sorrel, but in quantities that can cause gastrointestinal distress and even death, and in smaller, regular doses, they assist the body's formation of kidney stones. Do not eat rhubarb root, either, though it *is* a lovely peach color when disinterred and sliced open. Unless you seek relief for stopped bowels, in which case eat the root—a remedy recommended for thousands of years in China—but use moderation. As a medicine, rhubarb was once considered "a sheer delight," as Clifford M. Foust writes in *Rhubarb: The Wondrous Drug*, because it eliminated unbalanced humors without "increasing the pain and misery of the afflicted," unlike more drastic purgatives.

Native to China and Russia, rhubarb has medicinal properties that were first recorded in the oldest Chinese materia medica. It was coveted as a miracle drug by ancient Greeks, Mediterranean Europeans of the Middle Ages, and early modern Europeans, though no European at any of these times had any idea what rhubarb looked like, much less how to grow it. By seed? Cuttings? Grafts? The English name is rooted in *barbarum*, referring to the barbarian lands where rhubarb supposedly grew, and "Rha," the Greek name for the Volga, Europe's largest river. "Rhubarb": a goofy sort of word that would be adopted in the golden age of talkies to describe the chatter of background actors and used by Gertrude Stein before that, during the age of silent cinema, as she wrote her own glossolaliac glossary of difficult fruits, the "Food" section of *Tender Buttons*, where rhubarb is "susan not susan not seat in bunch toys not wild and laughable not in little places not in neglect and vegetable not in fold coal age not please."

Around 1770, someone (name unreported) smuggled rhubarb seeds into Britain from Russia in a diplomatic bag. Other sources report that an Isaac Oldacre, gardener to Sir Joseph Banks, brought the seed to England from where he was working in Russia, and that this is the cultivar that the nurseryman Joseph Myatt brought to Covent Garden Market in 1808 or 1809 or 1810, possibly the first time rhubarb was sold as a food. Only three of his five bundles attracted buyers. To customers who complained of a medicinal taste he recommended they cook the stalks in sugar. The following season, he planted an acre of the plant, gambling that late-winter/early-spring customers would choose rhubarb instead of last season's sad old apples.

He was right. By 1840, Myatt's son was tending twenty acres of rhubarb and shipping them to Covent Garden by the wagonload. Clifford M. Foust draws a causal relationship between rhubarb's popularity as a medicine and its turn as a dessert: "Had there not been two centuries of fascination with rhubarb as a cathartic, and repeated and widespread efforts to obtain the plant in order to accommodate it to Europe, it seems highly unlikely that it would have been 'discovered' as an attractive edible and have received the kind of fashionable attention that the drug had had for a century."

The earliest recorded culinary rhubarb recipes appeared around 1730 in the North American colonies, England, and France, including advice from the superintendent of exotics at Versailles for how to make rhubarb marmalade. The recipes were mostly tarts and jams made from rhubarb stalks, the same sweet use rhubarb is put to today. The stalks do not share the medicinal power of their roots or the poison of their leaves. They are sour or not very sour, but they are never, without sugar, sweet.

Except once, when I asked a crowd to eat raw rhubarb so I could take photos of their faces. Then the rhubarb I pulled from a friend's garden foiled me by being sweet and melonlike. I'd picked it from a shady, moist patch; perhaps I should have chosen from one fully rooted in the sun's glare. As the crowd chewed the cud of their raw rhubarb, their faces were confused and calm, replicas of how their tongues must have felt, having expected a certain amount of trouble.

As a child, my mother played this rhubarb game with her siblings: Whoever makes a face first loses. The winner is the child who chews longest without breaking the mask. This wasn't just a test of the child's endurance against sour, but also a test of the child's endurance against the texture of raw rhubarb. How celery-like fibers string themselves around the incisors and molars and make a gagging clump.

My mother remembers being big enough to reach the stove on a step stool to make formula for her sisters and brothers, the same stove we wipe down sixty years later to attract new owners after my grandmother Phyllis dies in the middle of Lent. Her rings had worn circular tracks through the numbers on the range dials where she'd gripped and turned them. These cannot be wiped away. We find hard candy in every drawer of her kitchen and each pocket of her purse, plus a note to herself, jumbled with her garage-door opener and two sticks of Trident gum. "Restless Hyper / feel very up tight," it says. At the end of her life, going blind, she would submit to injections in her eyeballs but refused medication to ease her anxiety. The last time I saw Phyllis alive, she seemed ecstatic that I'd come, then relieved when I left.

My mother can remember the ingredients of the formula she made her siblings, but not the recipe: milk, water, corn syrup, heat. She was first trusted to make this around age five, using the same sweet milk she drank in the time before time, when rhubarb games and most siblings hadn't been invented. When she grew a little older, she'd find Phyllis crying alone in the middle of the day. Remembers being slapped out of the room, confused by what she didn't understand—that she couldn't comfort her mother, that sometimes there's no comfort in children. My mother loses control of her voice when she gets to the part about comfort. "It's all I wanted," she says. Now that her mother is beyond comfort, she feels lighter, she confesses. Relieved of the weight of that want.

Of all fruits, I've heard the most ridiculous lies about rhubarb. That archaeologists can identify ancient Native American village sites in the Cascade Range by identifying old rhubarb patches. That thirty-seven people died after a World War II victory garden pamphlet advised eating the leaves. To find the best rhubarb-growing land in the world, go to the Mount Rainier Valley and follow volcano evacuation-route signs, but backward, toward the mountain. When the volcano erupts, first to go will be the Washington State rhubarb industry.

This is not a lie, but a secret: my grandmother once threw my mother against the kitchen wall so hard that she knocked a plaster cast of my mother's hand down from its nail and onto my mother's head, where it shattered. The first time my mother tells me this story, I forget it. She tells me again, then again, and each time it surprises me as if I've never heard it before, until an image roots—my grandmother sitting on her bed with her back to the door, her spine a raised runnel beneath her dress as she holds her head and cries, my mother walking quietly into the room, careful not to startle my grandmother, my mother reaching for her mother—.

I don't know why my grandmother was so sour. "Self-preservation," my mother has said. I know her father pulled her from school after the eighth grade; that at seventeen she pledged to marry a man her father wouldn't let her marry because they were, he said, cousins; that a few months later she married a different man, my grandfather, who was perfect and beloved by all; that Phyllis was estranged from her parents for reasons she never told her children. I know that from age twenty-two to thirty-two she was pregnant nearly every day. I know she loved her children, who never stopped trying to take care of her, even when she would not tolerate their help. She had an artistic eye, my grandfather's sister said. You could tell by how she decorated the house.

Rhubarb does not thrive in bright, hot sun. It bolts, sending strange bulbous rods from the dirt to burst into frothy, almost cruciferous-looking blooms. "Forced" rhubarb is grown under a ceramic cap or in a wooden box or kept in the dark at warmer temperatures by some other strategy, often in a greenhouse, sometimes lit only by a candle, which coaxes rhubarb from the ground from Christmas through Easter and makes the plants produce pale-pink stalks prized for their delicacy and sweetness, like the rhubarb of my failed game, too mellow for anyone to re-enact my mother's childhood feats of endurance.

To extend the harvest, pick the flowers. Take the flower's stalk in both hands and pull like you mean it. The stalk will be hollow, like a flute. Trim the ragged bottom so it is neat, green, firm, and fleshy. Put the rhubarb flowers in water, in a vase.

Pickled Rhubarb

Rhubarb stalks are too delicate to withstand the heat of processing for shelf-stable storage, so this is a refrigerator pickle. Use fresh rhubarb, preferably the thin, intense stalks of a struggling plant.

Yield: 16 ounces

1 cup champagne vinegar

¼ cup granulated sugar

¼ cup maple syrup

10 ounces fresh rhubarb (1½ to 2 cups, sliced according to instructions below)

4 garlic cloves

¼ teaspoon mustard seeds

¼ teaspoon coriander seeds

¼ teaspoon pink peppercorns

Wash and dry four 4-ounce glass jars and their lids. In a small saucepan, combine the vinegar, sugar, and maple syrup. Bring to a boil, stir to dissolve the sugar, then set aside the brine.

Cut off and discard the ends of the rhubarb, then slice the stalk into ¼-inch slices. If the rhubarb is very thick, cut the stalk in half lengthwise before making the ¼-inch slices.

In each glass jar, place one garlic clove and top off with rhubarb slices. Distribute the spices between the jars.

Pour the brine over the rhubarb, leaving a ¼ inch of headspace. Cover with the lids and refrigerate. It keeps for months, but is best in the first month, and good almost immediately—wait just twenty-four hours before serving.

Rhubarb Root Remedies for Stomach and Bowel Complaints

"In large doses," writes Maud Grieve in *A Modern Herbal*, "rhubarb powder acts as a simple and safe purgative, being regarded as one of the most valuable remedies we possess . . ." The same roots that produce the stalks we eat in pie can be used to make a stomachic to bind bowels, but for a good purgative that Grieve says effects "a brisk, healthy purge," choose Turkey rhubarb, a plant native to China and a relative of culinary rhubarb. All Turkey rhubarb was historically grown in China; its various names (Turkey rhubarb, Russian rhubarb, East Indian rhubarb) are related to the trade route by which they reached Europe, not where the roots were actually grown. At its full height, Turkey rhubarb towers over the tallest humans. When it isn't grown for medicine, it is grown as an imposing ornamental with broad, sharp leaves and dramatic panicles of white or pink flowers. The root must be gathered from plants that are at least six years old, Grieve insists. Dig them in October, then peel them, cut them in half, and hang them in chunks to dry. Depending on the quantity, they either produce constipation or end it.

Culinary rhubarb can also be harvested for its root, and anyway should be dug up and divided every four or five years to increase stalk production. Grieve recommends culinary-rhubarb root as a superior means to stop diarrhea, better than Turkey rhubarb.

To find Turkey rhubarb today, visit an herbal shop, where it will likely be sold by the ounce for no more than most loose-leaf teas. It will look like barky gravel and smell like old cinnamon with a hint of yellow mustard. To find culinary-rhubarb roots, check the nearest plant nursery.

Do not eat the leaves of culinary rhubarb or Turkey rhubarb. Depending on your size and physical condition, they can be poisonous.

John B. Lust's *The Herb Book* says prolonged use of this medicine is not advisable, because it "aggravates any tendency toward chronic constipation," but as an occasional medicine rhubarb root is an old and favored remedy. Pregnant and nursing women shouldn't use it, because rhubarb root may stimulate the uterus. Before using this medicine, consult an herbalist.

> ½ ounce Turkey-rhubarb root
> Water

Cover the rhubarb root in water, and soak overnight. To loosen bowels, take 1 tablespoon of the liquid rhubarb-root extract two or three times a day. To bind them, take ¼ teaspoon of rootstock in ½ cup clean water.

S: Sugarcane

Saccharum officinarum
Poaceae (grass) family

Sugarcane originated in New Guinea, was mentioned in Dioscorides' *De Materia Medica* as "a kind of concreted honey," was first processed in India, is said to have followed the Koran and arrived in Europe in the twelfth century, was carried to the New World by Columbus, and became the first industrial-scale slave-plantation crop in the mid-seventeenth century, in Barbados. The first time sugar appeared in written records, it appeared as medicine. From ancient Greece and Persia to pre–Industrial Revolution England, sugar was initially used to preserve medicine, which happened to make medicine more palatable (and still does—think Nyquil capsules and vitamin gummies) before becoming, itself, a kind of medicine one can eat. I mean that literally but also metaphorically, as the substance usually responsible for the emotional sweetness of sweets, the nostalgia that rustic pies and layer cakes trade on, balm for the mother-sick soul.

When sugar was still rare in medieval and early-modern Europe, that rarity helped make it more attractive as a remedy. Medieval prescriptions called for crushed pearls, amethysts, and sugar—cures only the rich could afford, whose only certain effect would have been the satisfaction of spending so much on one's health. Sugar's healthful properties have been questioned since the eighteenth century ("It could make ladies too fat," writes Dr. Frederick Slare around 1708—though he did think this was sugar's only defect, and otherwise it was "a veritable cure-all"), but no serious opposition to it as food or medicine arose until the late twentieth century. Today, it's the villain of clean eating, and I have more than once tried to give pie to a sick person who refused it because sugar was "killing" her. In 2016, the average American consumed ninety-four grams of sugar a day, or 358 calories, often from sugar added to sodas and juices and processed foods, nearly one-fifth of our total suggested calories, almost twice the daily limit the CDC recommends.

During the Middle Ages, sugar was worth as much as gold in Europe, but the price sank as its popularity and production rose during the industrial age, transforming it from a royal luxury in Elizabethan times (when a visitor to Elizabeth's court remarked on her black teeth) to a gauche staple of factory workers, "the first mass-produced exotic necessity of a proletarian working class," as Sidney Mintz describes it in his seminal book-length study *Sweetness and Power: The Place of Sugar in Modern History.* Those who couldn't afford fresh fruit ate sugary jam on bread and stirred sugar into tea, another stimulant that, with tobacco, broke the drudgery of the day. The goods English factory workers produced were shipped back to Barbados, where land deforested for sugar plantations was too valuable to plant with food crops that could have fed the local population, creating a trade dependency we now recognize as endemic to colonialism and ruinous to indigenous diets.

Once cut, cane that is left in the field ferments within a day. This created a logistical problem that European industrialists solved with white indentured servants and enslaved Africans, and then just enslaved Africans and their descendants, who carried out specialized work at all segments of the refining process, a division of labor not yet seen on this scale, with this intensity, and this propensity to destroy the body of the worker. From the time an enslaved person arrived in Barbados, his or her average life expectancy was seven years. Any pleasure we receive from processed sugar descends from that horror.

In 2014, the artist Kara Walker installed a giant sugar sculpture in a sugar refinery in Williamsburg, Brooklyn, and called it *A Subtlety, or the Marvelous Sugar Baby: an Homage to the unpaid and overworked Artisans who have refined our Sweet tastes from the cane fields to the Kitchens of the New World on the Occasion of the demolition of the Domino Sugar Refining Plant*. I began my participation in this event by taking my place in a long line of strangers who wanted to see inside the factory. When it was my turn, I walked through a large, dark opening that promised relief from the sun, like the entrance to a cinema.

Inside the plant, no one was working. Or—those who were working looked exactly like those who were not, except that those who were working were mostly rooted in place, waiting to answer should anyone ask about what they were seeing. As we entered the main room, a soaring space whose beams and walls were rimed with old molasses, the crowd dispersed in urgent flurries. I could have approached *The Marvelous Sugar Baby* or one of the *Banana Boys*. I could have inspected a wall. What I chose to see first is probably not significant. This was a free spectacle, and there was time to see everything.

Walker and her team constructed *The Marvelous Sugar Baby* of sculpted foam and eighty tons of granulated sugar. They made her almost as tall as the factory ceiling, and naked except for a headscarf. She was crouched like a sphinx, with a fathomless gaze, and *Sphinx* or *The Sugar Sphinx* is what some people called her. The thirteen *Banana Boys* wore sarongs and were as tall as my hip, maybe, or my chest. They were made of sugar, molasses, and resin and posed as if for work, with baskets to carry sugarcane in their arms or on their heads, smiling. Melting, too. Some had lost baskets, some had lost arms. Some drowned in puddles of themselves that we skirted as we approached for a closer look. When I retreated from a *Banana Boy* whose skin wept syrup, I had to pry my sandals from the sugary floor.

The mess was confusing. I wanted to mop it up. I wanted to sink my hands into the soft, sweet side of a *Banana Boy*, or break a chunk from the flank of the *Sphinx* and eat her.

Others wanted to turn their backs to the *Sphinx* and hold their hands up as if to cup her breasts, or approach her rear and mime cunnilingus, all while asking friends to take their photos. Others wanted to destroy the people who struck vulgar poses. Or wonder, or weep, or rage, or write an article. It surprised me, how I responded to the disintegration of that room. It surprised me how much other people were in the way of—inserted themselves into—my response. No matter how people behaved or what they captured with their cameras, they were likely to post it on Instagram, where, even after *The Marvelous Sugar Baby* and *Banana Boys* and Domino Sugar Refinery were torn down, they live a second life.

For the short time in the summer of 2014 when I was a live body in the room with these sculptures, I thought of childish things. Of Roald Dahl's *Charlie and the Chocolate Factory*, particularly the "Pure Imagination" number in the Gene Wilder movie adaptation. How, as children and parents gaze at Wonka's sugar garden, their wide-eyed wonder transforms into sublime squints of privacy. Violet closing her eyes to gnaw on her gummy bear; Veruca focused on smashing candy pumpkins; Mike Teavee's mom dipping bare fingers into candy mushroom cream, then abandoning propriety as she smears the cream into her mouth; Augustus Gloop beside the chocolate river (and himself), scooping liquid chocolate into his maw, so rapt he falls in and contaminates Wonka's pristine flow. Presented with a spectacle more like the one I was experiencing at the Domino factory than anything either of us had previously experienced, Master Gloop stuffs himself, sates himself, overbalances his body with thirst.

I thought of white-and-pink sugar packets stolen from the pizza parlors of my childhood and secreted to the bathroom, where I'd lock the stall door, rip the packets open, and pour their contents onto my tongue. In my spit, the sugar converted from a particulate into a pure liquid, and I was relieved for a moment from wanting before the wanting returned. It was crude to eat this way, and I knew it.

In the real-life sweet factory, none of us were ever alone with *The Marvelous Sugar Baby*. To take a photo without a person in the frame, I had to narrow my perspective, choose a particular detail of the sculpture, and draw as close as I could. What did the iPhone see? I made it frame the brown hand, the brown basket, the smile, the eye. The white paw, the white headscarf, the nose and mouth, the vulva. Better without the flash. In photos, the sculptures were more striking than the living people around them. They glistened and gleamed, their sugar lit by natural and artificial light. Every camera captured them easily, light sugar or dark sugar, against the shade of the room.

Sugarcane is best right after it's cut, and ferments within twenty-four hours. Before it can be eaten, the peel must be removed and the pith inside cut into bite-sized portions, from which a sweet, sticky liquid seeps. The part we eat is not the fruit, which you could easily conclude from a quick close-read of sugar*cane*'s name. Nor do we casually confuse refined sugar with what we call fruit. But we *do* define fruit by its adjacency to sugar—its sweetness. In true fresh fruits, that sweetness is from fructose, not sucrose, but the taste we crave is basically the same. Cucumbers and tomatoes, both fruits, are called vegetables because they aren't very sweet. Rhubarb, a vegetable, is lumped in with fruits because we eat it with sugar.

Sugar makes fruit more palatable, and also makes it edible for longer periods of time. In sugar we preserve, and with sugar we make preserves. On a material level, fruit isn't meant to survive the season, but sugar—which in its unrefined state is even more perishable than fruit—prevents fruit from going "off" or "bad." The words we use to describe sugarcane's action on fruit have a shade of immortality and judgment to them: In sugar, fruit won't perish. In sugar, fruit won't spoil.

The literal fruit this plant bears is, as it is for all grasses, thin and stuck to the seed coat. It's called a caryopsis (other examples are wheat and corn) and generally not eaten or used to propagate sugarcane, which grows better from stem cuttings, also called stetts. Sugarcane's figurative fruits, musically speaking (so I may narrow its range to a digestible size), run from raunchy hit singles like Trick Daddy's "Sugar (Gimme Some)" and Def Leppard's hair-anthem "Pour Some Sugar on Me" to Billie Holiday's beautiful and horrifying "Strange Fruit," with its image of lynched "black bodies swinging in the southern breeze," a "fruit" of New World slavery, the economic and social system that fueled—that was fueled by—Europe's sweet tooth.

By the fifteenth century and probably before, English royalty were familiar with fruit preserved in sugar, but fruit had a bad reputation among the working classes. Before jam became commonplace as a cheap calorie provider, people had shunned fruit for centuries because Galen-influenced ideas about health blamed consumption of fruit for all sorts of ill effects. By the nineteenth century, thanks to sugar, fruit had been transformed from poison (or at least *not good for you*) to what topped a workingman's bread when he couldn't afford butter.

Boiling fruit with sugar has two goals. First, it evaporates water; second, the sugar that replaces that water forms a set, creating a pleasing texture and trapping what moisture remains in a matrix of sugar, which deters its ability to house bacteria. Adding lemon juice increases the acidity already present in many fruits, preventing the basic-anaerobic environment that brews botulism. Of course, preserves still spoil. Over time, they darken, crystallize, toughen, and ferment; any jam that's been opened will eventually mold. But with suitably acidic jam, you can use your eyes and nose to tell if something's not right, and you don't need to worry about invisible bacterial killers.

More sugar makes an easier set, as with a raspberry jam that, though the fruit is poor in pectin, firms up just fine if the ratio of sugar to raspberries is 1:1. Below 65 percent sugar, the jam isn't as shelf-stable, advice I was given by Rebecca Staffel, the onetime owner of Seattle's Deluxe Foods, who offered a basic recipe of a thousand grams of fruit to 650 grams of sugar when I asked her the secret to going off-recipe while making jam. Kevin West, the author of *Saving the Season*, sometimes calls for only a half-pound of sugar to a pound of fruit, and I can attest to the high quality and life span of his jam recipes—though not their potential legality in the marketplace. Should one want to make money on one's experimental homemade jams using a Washington State Cottage Food Permit license, this ratio of sugar is insufficient. The state requires a pound of fruit to 1.2 pounds of sugar. Which means that, if I want to make and legally sell an intensely fruity, low-sugar jam, I'll need a commercial kitchen with all the attendant licenses.

It's a losing business plan, "impossible to make money at," Rebecca said, even though the product can sit on a shelf for long periods of time without spoiling. Fruit's expensive, sure, but I blame the way we think about preserves. They're the sort of thing people believe they can make cheaper themselves, don't always value as a luxury item, or—and this depends on the packaging—fear, if the jam appears homemade. "I'm afraid I'll get botulism," one friend said when I gave her a jar of my own. "Not that I think you're trying to poison me."

Finding actual sugarcane in twenty-first-century Washington State isn't easy. My cane cannot be anything like it's supposed to be, given its long journey and subsequent condition, poking from a utility bucket beside a banquet of grapes at my local grocery. It does not ooze and looks impossible to peel. There's a spot of red-and-blue glitter on one end of it. The other end has a split of about six inches, where I can see the fibrous inside and some fuzzy gray spots that appear to be mold. It looks, more than anything, like abused bamboo. I buy it anyway.

I can't smell the sugarcane. I can see, if I peer into the split on one end, that the center is woody, porous, like the middle of a tooth, or how I imagine the middle of a tooth to be. Though my kitchen cleaver is sharp, it is not heavy enough to break through my sugarcane's exterior. Instead, I start by hand on the end, where the outer hull has already split, muscling the cane open by tugging the split farther apart. The cane lets me extend the split six inches and then refuses, but that's enough. Now I can see a spongy core whose fibers suspend a whitish, dry, dimly sparkling substance I dig out with a steak knife.

"You need strong teeth, strong jaws, to chew sugarcane," writes the poet and performance artist Shailja Patel as part of *Creative Time Reports'* online presentation of *A Subtlety*. "The sticks are sometimes too thick to fit comfortably in the mouth," she writes. "You get purchase on a ridge of juicy, crisp fiber, arrange the fibers crosswise against the blades of your incisors. You position your tongue and the roof of your mouth out of reach of knife-sharp corners. You bite against the grain. If you bite parallel to the fibers, you end up splitting the cane, smaller and smaller sticks in your mouth, saliva running, no juice. But when you get the crunch just right and the surge and gush of sweet liquid—aaaah."

I appreciate the second person here, how I can read this as a recipe for my own chewing. But there's nothing Patel's recipe can do about my sugarcane's being too old. My bite is a small, dry piece. I chew it without precision. I can suck the sweet out of this stuff, but it isn't like juice. Anything above the length I've just opened, the part that had already split, is an angry red. Beautiful, but like an infection. Like mold has made it glow.

With few exceptions, sugar production has not fallen in five centuries. No known culture uniformly regards sweetness as unpleasant or rejects sugar outright. No country has rid itself of sugar once it has been introduced. Every green plant feeds itself with sugars made from photosynthesizing carbon dioxide and water.

Sugarcane and sugar beets have the highest natural concentration of sucrose. The sugar we pour from a C&H or Domino or Tate & Lyle bag (all owned by the same parent company, ASR Group) might be a mixture of cane and beet, spun in a centrifuge to remove molasses and dried into fine, regular, snow-white crystals.

To refine sugar, you must crush the cane; extract the juice; purify, clarify, and evaporate the juice (sometimes using the ash of animal bones); crystallize the juice by extracting the moisture; then spin the crystals very quickly, so centrifugal force may separate them from molasses. Then the crystals must be dried and bagged, the molasses captured and bottled.

Brown sugar, light or dark, is white granulated sugar with molasses added back in. Cane sugar is like granulated sugar except definitely from sugarcane (not sugar beets) and with a little molasses left, so more caramel flavor. In this historical moment, cane sugar is considered the new "fancy sugar." Caster sugar is superfine granulated sugar. Demerara sugar is chunky, dry, with some molasses flavor, making it caramel and toffeelike. Granulated or white sugar (formerly known as "fancy sugar" or "company sugar") is superrefined, with all the molasses spun out. It is dry, plain, and sweet, and can be made from sugar beets. "Invert sugar" is a liquid sugar that has been processed to split glucose from fructose and balance these components in equal parts. It is used for industrial purposes to retain moisture or slow down crystallization. To make it at home, boil sugar, water, and lemon juice gently. Jaggery is concentrated unrefined sugarcane syrup and date palm syrup, crystallized in blocks. Molasses is the by-product of sugar making (or, seen from a more enthusiastic perspective, the nutrient-rich syrup extracted while processing cane juice). Muscovado sugar is a fancy brown sugar that, unlike conventional brown sugar, has not had the molasses removed. Panela, like jaggery, is made of boiled and evaporated sugarcane juice and generally sold in blocks. Panela resembles muscovado but is harder; it is served by scraping or chiseling off sections. Powdered sugar or confectioner's sugar is granulated sugar that has been crushed into a powder. Sucanat is extracted from sugarcane juice and minimally processed; it has a deep molasses flavor. Turbinado is chunkier than white granulated sugar and a bit more brown, not moist like muscovado, and with a little molasses flavor. Turbinado crystals are a little smaller and browner than demerara crystals.

Kara Walker's *Homage to the unpaid and overworked Artisans who have refined our Sweet tastes* was inspired by medieval subtleties, figures sculpted from malleable sugar (often marzipan or sweetened plant gums) and served between courses at royal dinners in order to send a message to the devourer. "By eating these strange symbols of his power, his guests validated that power," writes Mintz, whose tattered and notated pages we see Walker read in the Art21 film about her influences. Subtleties also resemble maquettes, the tiny replicas of large-scale public works that sculptors make to test out an idea, and may keep to commemorate work that cannot fit in a private room. I'm not sure if Walker had maquettes in mind, but their small scale reminds me of the medieval subtleties that inspired her. What if *The Marvelous Sugar Baby* is the extreme logical conclusion of maquette-sized subtleties meant to reinforce the power of kings whose scions fattened themselves on the stolen lives of Walker's ancestors? We could then say the kings' subtleties were just for practice and Walker's *Marvelous Sugar Baby* is the real thing. "Initially, the displays were important simply because they were both pretty and edible," Mintz writes of subtleties. "But over time, the creative impulses of the confectioners were pressed into essentially political symbolic service." I think it's notable that visitors to *A Subtlety* were not in any way invited to eat something, sugared or otherwise. Modern takes on edible subtleties would have been easy to include—sugar candy is cheaper than ever—but this is not how the art asked us to participate.

"Sugar is even more important for what it reveals than for what it does," Mintz writes.

What did the sugar of *A Subtlety* reveal?

Some looked at the Sphinx and the *Banana Boys*, amazed. Some preserved the experience with photos. Some took lewd photos. Some thought this behavior was symbolic violence by white people against people of color. Some organized groups of people of color to enter the space together, to view the art from the safety of numbers and draw attention to how often art spaces are white spaces. This spectacle of sugar and decay and taking photos and posing for the camera and imposing ourselves on other people's pictures/experiences/feelings created the dynamic that attracts me to Kara Walker's work: the art was crowded (figuratively via composition and literally via human bodies), the crowd did not agree with itself, and the art did not suggest a correct way for us to experience it. That's a job crowd members found themselves taking on largely in response to how other members' behavior hurt or oppressed them.

Walker's work is often unnerving. Her imagery deals graphically (and beautifully, which is even more uncomfortable) with the enslavement, rape, and murder of Black people. *The Marvelous Sugar Baby* was a memorial to Black ancestors and a challenge to white supremacy, *and* it invited all viewers to react as they would and record that reaction, drawing their reverent and awestruck and bewildering and hurtful responses out in the open and putting them, too, on display. "This wasn't just going to be about, 'We're all [messed] up and we're all gonna die and we're all trafficking in other people's bodies,'" Walker told the *Los Angeles Times* in the fall of 2014, after the *Sugar Baby*'s demolition. "We were also building something fantastic. So I had to go with that spirit: 'We're going to change the world and we're going to do it like this!' But the gist of the piece was that it wouldn't be rebuilt again, that it would never happen again. It was ephemeral. You build these monuments, but they're really castles in the sand. It's like sugar. It evaporates and goes away."

When I stepped away from the *Banana Boys*, my tracks disappeared into the tracks of those around me. I was a small woman in a big crowd—a tourist, a worshipper, a vandal. I wished, at the time, that something would create a unified point within the experience so I could be relieved from the overwhelming wash of it, like what might happen if a member of the crowd broke into song, and we could listen or join in, all these photo-snapping, awe-full people.

On that day in the sugar plant, no one sang, and I did not taste anything sweet. Later, as I was undressing for a shower, sugar-smell steamed from my clothes, then rose from my hair in one strong puff, mixed with my scalp's animal scent. I smelled this most intensely right before I did what it was time to do: duck my head beneath the hot water.

Cane Sugar Scrub

It took six months to get my homework just right: exfoliate daily in the four weeks between appointments, arrive hydrated (no whiskey the night before!), having forgone all moisturizer for twenty-four hours. Failure to comply results in painful, incomplete sugaring that my aesthetician cannot do anything about, a failure she'll narrate if requested so I might do better next time.

To sugar my legs, Traci dips her hand into a heated jar of thick amber liquid and twirls her wrist a little as she brings it up to form into a ball. "I didn't realize until I tried to teach someone to sugar how hard this is," she says. "You have to have a knack for it." If you don't, she can tell immediately. You'd have sugar everywhere, all over yourself, strings of it in your hair, on your arms, your face. It wouldn't even take that level of disaster to indicate you should stick to waxing. "It's hard to get it in a ball, like this." Traci demonstrates, dipping two gloved fingers into a warm jar of cooked sugar. "If you don't pick this up just right, it gets it all over your palms and wrists instead of forming a ball. Once that happens, it's all over."

The substance she's using to remove the hair from my legs is basically candy—just sugar, water, and lemon juice cooked to what candy makers would call the soft-ball stage, no preservatives, made in small batches, viscous enough to form a solid shape but liquid enough when heated to spread down my shins and calves. As Traci works, spreading the sugar in a thin layer, then ripping it up and rolling it back into a ball, the sugar turns from dark translucent amber to opaque yellow. "That's dead skin," she says. I can see my hair floating in it. Once the sugar ball is light yellow and littered with a layer of me that makes it the consistency of silly putty, it stops grabbing hair. To throw the ball away, Traci pulls her glove over it, then layers the other glove over that and tosses it all in the trash. She tries to use two balls, but if I've been bad about exfoliating or if it's winter, she has to use three or four.

It's uncommon to sugar legs, Traci says. The results aren't as Barbie-smooth as shaving, and they guarantee at least ten days of wearing your legs hairy before the hair has grown in enough for the sugar to catch hold and yank it back out. I sugar because I hate shaving, and I like subverting beauty standards about 30 percent of the time, and there's something fun and clean about pulling all that hair out. It's a satisfying hurt. Most people who sugar, sugar their bikini lines. Some sugar their entire vulvas, a process a friend describes as excruciating but "worth it." It makes sex feel better, she says. I can't bring myself to try it. Too much pain and exposure for my taste. Too much grooming.

Traci's dog likes to chew on her clients' socks—I hide mine under my pile of pants and coat while I'm being sugared—but isn't interested in the lumps of sugar in the garbage. While I balance on one leg to put my jeans on, she licks the ankle carrying all my weight.

To prepare the legs for sugaring, one must exfoliate consistently. Otherwise, as hair regrows, it gets trapped beneath the skin, causing red bumps and ingrown hairs. To make the scrub Traci recommends, use more sugar:

In a medium bowl, combine 2½ cups cane sugar (organic if you can get it, so the crystals are bigger—the point is to have crystals big enough to exfoliate but not big enough to scrape) with 1 cup sweet almond oil. Add scented essential oil to perfume the scrub, as much as you like. Pour the scrub into an airtight jar and leave the jar in the shower for easy access. To exfoliate skin, first soak yourself in hot water, then scoop some sugar scrub from the jar with your fingertips and rub it all over, every day. Remember to rinse.

Sugar Pills

For those who would test the placebo effect, a "medicine" (even in air quotes we will regard this as medicine as long as we write the word "medicine") that is reimagined with cane sugar.

Yield: 60 pills

60 empty gelatin pill capsules

Sugar of your choosing (consider color and texture over taste, how the sparkle or lack thereof, how the white or brown or somewhere in between, will add or detract from your belief in the sugar pills' salubrious powers)

1 empty pill bottle

Choose a condition you'd like to cure, a dream you'd like to have, a truth you want to believe, a quality you'd like to possess. Fill the gel capsules with sugar, then fill the pill bottle with the sugar pills. Label the bottle with this cure, dream, belief, or quality. Every morning, read this word as you unscrew the bottle and remove one pill. Swallow the pill as you'd swallow any pill. With water. With or without food. Or look at the pill and throw it away. You may notice a subtle improvement of your condition. Continue as needed.

If your condition deteriorates, cease taking this medicine and consult a doctor, life coach, witch, family member, or friend.

T: Thimbleberry

Rubus parviflorus
Rosaceae (rose) family

The description of thimbleberries in the *New Oxford American Dictionary* was written by someone who's never had a thimbleberry. They are, it says, a "juicy, somewhat tasteless fruit."

As anyone who has eaten thimbleberries knows, they have a rich, raspberry-like flavor that's more intense than one would expect, as if it's been concentrated by gentle heat. Though there is juice in a thimbleberry, it's not the sort one needs a napkin for. "My favorite of all the trail berries," more than one friend claims.

To find thimbleberries, look for a trailing vine with

large, soft leaves and no thorns, growing in shady, moist, low-elevation forests from California to Alaska, especially along hiking trails. Thimbleberry flowers, like blackberry flowers, emerge in spring papery white and backed by star-shaped sepals. Those flowers transform into berries that ripen in July and August, blushing from white to deep pink only a few at a time. Their magenta drupelets loosen from receptacles the way raspberries do, but they are much shallower and often wider than raspberries. Thimbleberry leaves are large and lobed like maple leaves. They're also fuzzy and soft to the touch, unlike any other *Rubus* species. They are, I've been told, the hiker's toilet paper of choice.

Like raspberries and blackberries, thimbleberries are not berries, but an aggregate fruit. We call them "berries" because they are small and berrylike. In English, thimble-berries are named for their resemblance to a sewing tool. In native languages their names include *sla′ka* (Upper Skagit), *la ′qa′ats* (Swinomish), and *taqa·′tcitł* (Quileute). According to Erna Gunther's *Ethnobotany of Western Washington*, the Makah make tea from *lūlūwa′ts* leaves for an antianemia tonic, and the Cowlitz dry and powder *kᵘku·′cnas* leaves and use them to prevent burn scars.

Unless picked with extreme care, thimbleberries tear and disintegrate. They do not keep well and are impossible to gather in large quantities. There is no commercial market for them. Nor do they appear as a primary ingredient in sweet or savory recipes, even in most guides to wild berries. Gunther's ethnobotanical records of the Northwest's native tribes don't mention culinary preparations, either, though that doesn't mean the berries weren't (and aren't) eaten fresh off the vine with delight, or dried and included in pemmican and other foods. My recipe tester for this chapter—a rare person with a thimbleberry bush in her own backyard, the only person I knew who could reasonably expect to have a dependable crop of thimbleberries—reported in September that her bushes didn't fruit. No one who knows thimbleberries will be surprised.

At the Willows Inn, the Pacific Northwest's answer to hyper-local Noma-inspired cuisine, foraged thimbleberries are the climax of an already memorable summer meal, served fresh and plain, luxurious because of the effort to find, pick, and transport a sufficient quantity of them. So understated, the dish shouts to fine-diners: All your life you could have had *this*, the very thing you've been missing.

Foraging is what you call berry picking when you do it because you can, not because you have to, my friend Maya points out. As with huckleberries, a fruit that's also free for the taking, art and commerce transform thimbleberries into a luxury experience out of reach for some of the people most familiar with them. These thimbleberries I call foraged, the same thimbleberries painstakingly sourced and served by the Willows—growing up, Maya just called them food. In the summer, she'd eat breakfast, then go outside and play in the woods all day, grazing from vines and trees. There was food at home, but not good enough to come home for. "Someone might think, 'Oh, that poor kid didn't have anything to eat,' but that's not how I saw it," she says. "I could go outside whenever I was hungry. *That's* fresh produce." When she ate thimbleberries right off the vine, it made her feel rich. "The necessity is a privilege," she says. One she does not romanticize now that she's achieved a measure of middle-class security, but one she misses.

Thimbleberries, salmonberries, huckleberries, and black-berries could help keep a lost hiker alive, or make an an-cestor's medicine, or feed a child who doesn't want to go home. I have always had plenty to eat and been generally averse to hiking (though, as I learn the names of plants and what they can do, I've changed). When I can find thimble-berries on the well-marked trails I prefer, they present me with the dilemma of someone who shares her path with too many people: If I don't pick this near-ripe berry now, some-one else will. If I pick the berry now, I'll ruin its chance to reach peak form. If I don't pick the berry, I won't know how it tastes. Should I pick the berry?

If I were not myself, I might be the sort of person who knows secret thimbleberry spots and suffers no anxiety that some human will pick them before I get there. But I am me, and rarely present at quite the right, ripe time. Even so, of this minor, local, special fruit, I already know more than the *New Oxford American Dictionary*. Thimbleberries are not juicy. They are not somewhat tasteless. They are the best of all the trail berries, if you can find them.

Thimbleberry Kvass

It is very difficult to source thimbleberries in the quantities required for jams or pies, and even hard to gather enough for a single mouthful. Should you find more than a handful of them and you don't eat them right on the spot (which you should), try this kvass. A fermented drink from Eastern Europe and Russia, kvass is half as sweet as Coca-Cola, and lightly carbonated, thanks to the fermentation action of naturally occurring yeasts on the berries and in your kitchen. Assuming you'll have as much trouble as I do finding thimbleberries, I've written the recipe to accommodate a small amount. All the measurements below can be increased or decreased to suit the quantity of thimbleberries you're able to find; just make sure your ratio of berries to sugar to water is the same. Fermentation takes three to five days.

This recipe is adapted from Sandor Ellix Katz's recipe for strawberry kvass in *Wild Fermentation*, a wonderful book that documents many traditional fermentation processes like this one.

Yield: not quite 1 cup

¼ pound (heaping ½ cup) just-picked thimbleberries
¾ cup water
1 tablespoon sugar

Rinse the thimbleberries, and place them in the bottom of a clean glass pint jar.

Gently heat the water, and stir in the sugar to dissolve, then remove from the heat to cool. Katz recommends dechlorinated water, but I haven't noticed a difference when I skip this step. Mostly, I'm lazy and I skip it.

Once the sugar water is cool, pour it over the berries. Katz says the vessel should only be half full, with plenty of room to stir.

Cover the pint jar with muslin or a thin tea towel, secure the cloth with a rubber band, and let it sit on the kitchen counter, but not in direct sun, for three to five days. Stir kvass "as often as you think of it, at least two to three times per day," to distribute fermentation activity and make sure all the fruit spends some time submerged. After a few days, the kvass will start to bubble. Katz says that if you stir more frequently the bubbles and fermentation will be more active. Look for a changed color, shrunken berries, and a strong aroma of August in Western woods (sans wildfire smoke, I hope). Strain the kvass and drink. It's especially delicious chilled and served on ice. Store any leftovers in the refrigerator.

Hiker's Toilet Paper

Identify thimbleberry leaves by their maple-leaf shape, fuzzy texture, and large size, ranging from four to eight inches across. Pick them from the vine and find a private place off the trail. You know what to do next.

U: Ume Plum

Prunus mume
Rosaceae (rose) family
Also known as Chinese plum,
Japanese apricot, mume, mei

I'm working outside, and the air smells like woodsmoke.
Which is nice at first.

It's the last day of summer, ninety-five degrees or will
be, bright and shadowless thanks to wildfires burning to
the east and west of where I sit. Five-day forecast: same.

Two days ago, Sam and I drove from Seattle to Spokane through Snoqualmie Pass, where a black billowing
column from the Cle Elum fire leaned over the freeway like
a close talker. At home, closer to the Montana wildfires,
the smoke is a thick, sourceless haze. Tiny particles of ash
bend white light to orange and make a full moon glow like

a cheddar wheel. This happened last year and two years before that. We're beginning to think it's a season: winter, spring, smoke, fall.

Yesterday, a man ran into the bonfire at Burning Man, and people can't tell if he did it to kill himself or if he did it because he was high. There's a small article about him in the national section of the newspaper, which Sam points out because he knows I loathe Burning Man. He wants to know why I hate it so much. Not because he actually wants to know, but because he thinks my discomfort is funny. In the mid-1980s, Sam was a Deadhead; I grew up thinking hippies were ridiculous. The joke's on me—now hipsters are ridiculous. Regardless, we do not share the same feelings about transcendent group experiences or psychedelic drugs.

"It's stupid," I say, as he expected me to say.

"But why?" he says.

I tell him it's a church with no judgment and no stakes, like astrology. I can see why such routes to transcendence would appeal, but, even having left the Catholic Church, I'm still Catholic. I still raise my eyebrows at religious (or religious-adjacent) experiences that don't engage with the possibility of punishment.

On the patio table where I'm writing, there are ashes, maybe from the sky but more likely from my stepdaughter's cigarettes. Jane knows she has the tobacco gene, just like Sam knew, but it's hard to give that knowledge serious belief when it's just a few cigs a day, when you still tell yourself you have a choice. That's how I talked to myself about my own habit, anyway, more of an inconvenience to other smokers than a threat to my health. I was a chipper, a bummer, a "Sorry, can I have one?" Not addicted, no way, not even when cigarettes stopped tasting good, which Sam says is crazy: "Cigarettes never, ever, ever stop tasting good." As for worrying about Jane's habit, one thing I know for sure is that a stepmother's Plan A should be to mind her own business.

Watching me smoke my twenty cigarettes a year made Sam want to start again. He's the only convincing reason I've ever found to quit completely.

He'll start smoking again in fifteen years, when he's seventy, he says. I make a dread-face. Hey, he says, I need this. It's the only bargain that worked.

That and hypnotism, which he didn't believe in until he fell out of his chair at the end of the session and, as of this writing, never smoked again.

Every now and then, I try to reopen negotiations with him, ask him to wait until he's seventy-five or eighty. He listens, promises nothing. He knows, from years of failure, that when it comes to cigarettes the only promises he can keep are the ones he's negotiated with himself.

On the benches that frame this patio, umeboshi are drying. I don't have the straw mats the recipes call for, and I don't have direct sunlight because of the smoke, so I put them on dehydrator trays and put those on top of metal baking sheets, thinking maybe the reflected heat will help. My instructions to make umeboshi are, roughly, pack ume plums in salt so they release their juice, weigh them down until they're submerged in salt-juice brine for three weeks, then drain them, leave them in the hot sun for three days, and finish by shutting them in airtight darkness to ferment for at least a year.

The first part of the recipe starts in mid-spring and finishes in early summer if you have real ume, which I don't. Having failed to find a local ume-seller, I called the Seattle Uwajimaya produce department in June, which had one more case of ume, but just barely. The fruit had turned yellow and was breaking out in brown spots. They wouldn't in good conscience ship them across the Cascades. I should have called in March.

From where I write at my patio table in the smoke, it is early September. The umeboshi drying around me are actually shiroboshi, made from unripe shiro plums, a sweet yellow variety I chose because they're similar to ume in size and easy to source in eastern Washington, and because they happened to be not quite ripe the early-July week when I went looking for an ume substitute. After their first ferment, from mid-July to early August, I sun-dried the shiro and packed them away as instructed. This is a second batch, made even later in the season, because I was anxious I'd messed up the first. I barely have time to sun-dry these shiro before fall hits hard, as it does around here. On top of that, umeboshi take a year to reach an acceptable state of fermentation and are best after five years, so any mistakes I make this summer won't be discovered or corrected for longer than it will take to write this book.

Ume plum blossoms unfurl in pink and white waves across Japan in January, also described as a "front," like a storm, and celebrated as Ume Matsuri, a sign that the most horrible part of the winter—my favorite part, the dramatic part—is over. Ume Matsuri is the festival before Hanami (blossom viewing) and Sakura (cherry-blossom season), and doesn't attract the kind of tourism that cherry blos-

soms do, for which maps of Japan are drawn not just with
the average dates of the season, but with predictions for the
coming year's blooming dates. "Ume blossoms are a sure
sign that winter is in full force. The new year has arrived,
and spring is just around the corner," write the editors of
Sukiya Living: The Journal of Japanese Gardening, in a series
of clauses that slip so quickly past the force of winter into
the springness to come that it feels to me like they're hiding
something. The trees are native to Japan, Taiwan, Korea,
and China, where Xie Xie (謝燮) wrote a poem during the
Chen dynasty (557–89) called "Early Plum" (早梅), one of
the only examples of his work that still survives:

> *Blossoms come early to greet spring*
> *Alone and not afraid of the cold*
> *Fearing if late in coming behind the others*
> *No one would be attracted*

I found a brief reference to this poem in a 2011 article in *Chronica Horticulturae*, but could not find further
information via English-language libraries or search en-
gines. My friend's dad, a retired Boeing executive formerly
based in Taipei and Beijing, researched Xie Xie in Chinese-
language sources, translated this poem for me, and drew
my attention to another place where ume can be found:
the logo of China Airlines, the national airline of Taiwan,
which in 1995 replaced the Taiwanese flag that had been
emblazoned on the tail assemblies of its planes with an illustration of a pink ume blossom. At the time, *The Journal
of Commerce and Commercial Bulletin* reported that the
move "generated much speculation among Taiwanese and

foreign observers" about "whether the new logo was adopted to appease China, which considers Taiwan as a renegade province." On its first flight, the new ume blossom decal fell off.

Ume fruit ripens in early summer. It is small and sour, full of citric acid and minerals such as manganese that are credited for its association with detoxification and germicide. Some books say the skin is green and some say it is yellow and some say that when the fruits ripen they blush slightly peach. Having not yet obtained a quantity of them for myself, I do not know the color of their raw flesh. I do not know if ume plums stay firm when ripe the way Italian plums do, or if they are more like Luther Burbank's Santa Rosa plum, soft and juicy. Since ume lends itself to pickles and processing and has the nickname "Japanese apricot," my guess would be that it's firm. As I've searched for ume in my own city, I've received two leads on sour green plums that homeowners can't identify and don't eat, and would I like them? Both turned out to be greengage plums, prized in Europe and delicious jammers. Not ume, but not a total failure.

I've been working outside for an hour now. My sinuses burn when I inhale, and I think I'm getting sunburned through the smoke. I always think I'm getting sunburned. I always am.

No ume where I live, but sweet shiro plums are everywhere. They're canary yellow, ripe in August, and appropriately sour if picked in July, a tip I gleaned from my mentor Lora Lea Misterly of Quillisascut Farm. She transforms unripe plums picked from pruned branches into "olives" by stuffing them in jars with vinegar or plain salt brine. By the end of summer, they become a crunchy, astringent treat that no one will mistake for real olives, but that make an interesting substitute on a cheese plate. Ripe shiro are bred for out-of-hand eating, the sort of plum you might buy at the farmers' market and enjoy in three bites while thinking about something else.

Umeboshi advice from the *Zohei Monogatari*, a Japanese military tale written in the sixteenth century, as quoted by Kōsai Matsumoto II in *Traditional Herbs for Natural Healing*: "When you get to be short of breath after working very hard, take Salted Plums from the provisions bag at your side and just look at it. You must never lick it. You must not eat it either, of course, for just looking at Salted Plums is enough to make you thirsty. While you are alive, you must value the Salted Plums. If you become thirsty looking at Salted Plums, skim the topmost layers off a pool of muddy water or drink the blood of dead bodies." According to tradition and folklore, umeboshi can be used as a detoxifier, an alkalizer, and a water purifier; when they accompany bento and onigiri, their salt content is said to help preserve rice and fish; and their minerals and acids contribute to a macrobiotic diet that could help me with my gut trouble if I just stuck to it long enough. I wonder if this belief is related to the belief in vitamin C, which ume contain in high quantities, as a cure-all.

Because there are few English-language single-subject books on ume and I like the bright-yellow cover, I buy Kōsai Matsumoto II's 1977 self-published manifesto of the healing powers of ume plums and reishi mushrooms.

Traditional Herbs for Natural Healing sounds like a number of herbal and holistic books currently in print, but I'm pretty sure this is the only title that also serves as a book-length advertisement for meitan, a granular plum extract developed by Matsumoto's father in the first half of the twentieth century that could be reconstituted in hot water and drunk like Emergen-C.

According to Matsumoto II's pitch, meitan plum extract combats radioactivity (this assertion paired with an illustration of a mushroom cloud), mental illness (illustration of a man strapped down with his head restrained by another man while he receives a forced injection of ume extract from a third man), stress (a businessman with an office view of jumbo jets), bad skin (a woman in a towel, inspecting herself in a mirror), hangovers (a whiskey bottle with a rocks glass), anemia during pregnancy (a woman in a rocker, knitting), plus applications for "Athlete's Foot, Snake Bites, and as a Unique Contraceptive" (a toilet/bidet with a flower on the seat). To use plum extract as a contraceptive, Matsumoto advises women to "take one teaspoonful and dilute it in ten times as much water. Put this solution in a douche and put half of it into your vagina while lying on your back before making love. Use the other half of it later to make sure of contraception. Plum extract kills sperm with its acidity and you can use it without worrying about harmful side-effects."

Oh, to have the faith of a salesman!

Actual ash is floating onto me and this notebook and joining the ash already on the table. The light is yellow, and the sun is pink. The salted shiso scattered between the shiroboshi look like clots of ash. The real ash looks like dandruff.

Eagle Creek is burning. Missoula is burning. Jolly Mountain is burning. My head hurts too much now for me to stay outside, and I shouldn't have blamed Jane for the mess on the table. Forgive me, Jane! It was less scary than global warming.

I gather my things but leave the plums. They still have two days to dry under the sun, after which they'll be brown and wrinkled, with a desiccated film on the outside that cracks under pressure, revealing a moist interior. I'm not worried about the ash that's still falling from the sky. It is a condition of this particular preparation of umeboshi.

Once I know enough about umeboshi to ask halfway-intelligent questions, I get in touch with my friend the musician Tomo Nakayama, to see what he remembers about growing up with umeboshi. "They're sour and overwhelming!" he says. "I have a taste for them now, but I remember wondering as a kid, 'Why do people eat this? It's gross.'" In his Japanese American home, umeboshi were always around, a hallmark of bento culture, chopped and laid in a circle in the center of rice to resemble the Japanese flag. "Think of umeboshi as a standard condiment," he says. "You don't usually eat more than one or two"—in part because of their strong flavor, but also because they take such a long time to prepare. Though he doesn't know what it's like in modern urban Japan, where his extended family can buy umeboshi in bulk in grocery stores, people in the country-side still prepare their own, and each family has its own way of doing so. "One of the classic or almost cliché scenes you'll see in a Japanese TV show or movie," Tomo says, "is, when Grandmother dies, the family will keep eating the umeboshi she pickled. Eventually, they'll run out. That's the moment when they really accept their grief." Did that happen when his grandmother died? "I feel like it did," Tomo says. "She was alive but she had Alzheimer's, so we hadn't been able to see or talk to her for a few years. We'd commune with her through her food and different recipes. I feel like my mom talked about that a lot—the things my grandma used to make and how we could still eat it, even after she died. But I can't remember if that's what actually happened or if that's just a scene I'm familiar with."

Umeboshi advice from modern recipes: Be careful not to tear the plum or include plums with bruised or perforated flesh. These imperfections are vulnerable to mold, which can spread to the whole batch. Pull the stems straight out from the plum so they pop free without grabbing the plum's flesh. Lower salt quantities increase the risk of mold but make a healthier product. Don't throw out a moldy batch. Just pick the mold off. "You'll know if it tastes funny," says Hiroko Asakura, the cofounder of Kamonegi in Seattle, one of four restaurants in the United States that serve handmade, hand-cut soba. "In Japan today," she says, "making umeboshi is a bit of a novelty. Like, 'Wow, you know how to make pickled plums? Can you show me?' It's a trend. People want to do these things their grandparents did. But umeboshi are definitely in every grocery store. You can even find them at a 7-Eleven."

Hiroko describes an ume paste that is not a condiment but a medicine, strong-tasting and dense, made by grating raw green ume and cooking it slowly into a dark-red molasses-like syrup. "This is not something you eat a lot of," she says. "I was told to gargle with it or use it to bring my immune system up." To cure an upset stomach, eat a spoonful of ume essence, or put it in hot water and sip. She's heard that this essence can help skin troubles and soothe stiff muscles. She says, "Think of it like apple-cider vinegar," a common household foodstuff used in the United States to address anything from stomach complaints to dirty sinks. "No other fruit has this medicinal quality that is used in so many different forms," Hiroko says. "It's so present in our lives, we don't even think of it as medicinal."

Prunus mume is closely related to apricots and is sometimes called Japanese apricot, but to make umeboshi at home I should not substitute apricots, says Makiko Itoh, the author of *The Just Bento Cookbook*. Nancy Singleton Hachisu, the author of *Preserving the Japanese Way*, is open to this practice but has not tried it. They agree on the method: I should layer salted leaves of shiso (also known as perilla; also called beefsteak plant by the Victorians, who used their serrated purple leaves to fill out flower beds) and plums (they use ripe ume, I use unripe shiro) in salt measured to create a 12 percent brine from the plum juice it will eventually extract (Hiroko advises using 20 percent salt to be sure the plums last all five years). I should layer the plums and salt and shiso in a crock bought for this purpose, or in a glass jar reused for this purpose, and in this crock I should press them under a weight of equal measure to the plums in a dark corner of my house for three weeks. Then I should strain off and save that liquid (a vinegar called "umesu") and sun-dry the plums for three days. They do not have to be consecutive days, Nancy Hachisu says. In Japan, Hiroko says, you'd ferment the ume during a June monsoon season, then dry them outside during the arid hot season that follows. "Now, because of climate change, it's more like July." Finally, I should pack the semidried plums in airtight containers to soften and ferment in the juice remaining to them, and leave them alone for half a decade.

It's a cocky recipe, I tell the neighbors, who don't understand why I'm leaving food outside in these conditions. It assumes we'll still be here in five years.

Hiroko also says putting plums in the sun is partly ritual. "You can make umeboshi without sun-drying them, but people say the sunlight is cleansing, that it will help the umeboshi last longer. Some people say sun-drying makes their color better." I can confirm this, I tell her. "Sometimes what you learn from your great-grandparents can be a myth and sometimes it can be true, you know?" She was told not to eat too many *tenjinsama*, the "little gods" hiding within a plum pit, because it would give her a stomachache. This was good advice. Like the pits of any stone fruit, ume plum kernels contain a little cyanide, which the body metabolizes in the gut. If you eat too many kernels, that's where you'll feel the trouble first.

The week after my day in the haze, after the weather cools and the smoke blows out and fall drops like a curtain, I'll dry a second batch of shiroboshi in my dehydrator inside, where the air conditioning will run day and night to filter the last of the air pollution from our house. These plums retain their yellow hue and seem less clearly layered in their dry-exterior/moist-interior-ness. I do not know if this matters, except that the outdoor-dried umeboshi look like the photographs in the recipes I'm following, and these indoor versions don't.

The brine is salty, shiso-flavored. I suck that off my fingers. Hiroko tells me I can use it to pickle cucumbers and turnips, or pour it over rice, or use it with sweet shoyu to marinate small fish, especially sardines. I'll start with pickles, the easiest thing, and put my three cups of umesu in the back of the fridge to chill until I need it—though it's so salty, I could probably store it at room temperature. I have been advised by other sources to use umesu on cabbage, as a salad dressing, or to take a shot of it to wake myself up in the morning. Umeboshi, too, are a swear-by morning tonic for the sort of people who like something bracing before 9:00 a.m.

I have five years to eat other people's umeboshi before it's time to figure out what to do with all of mine. A schedule that won't fit the production of this book, not that ume cares. Jane will be too old for her dad's health insurance by then. I hope Sam and I will still be here, in this house, taping our windows shut for fire season, not yet smoking, but getting closer.

Ume/Shiroboshi

In the summer of 2017, I conducted a four-part experiment to see which corners I could cut in Nancy Hachisu's and Makiko Itoh's umeboshi recipes. The biggest corner was the one most successfully cut: if you can't find ripe ume, unripe shiro obtained around the time sour cherries are ripe deliver the necessary sour pucker and small size. What didn't work: oven-drying made goopy, sodden shiroboshi, as did fermenting ripe shiro plums, whose sweetness poorly balanced their salt. What sort of worked: drying the unripe shiro in a dehydrator is an acceptable substitute for the sun, but these shiroboshi were paler in hue. Two years later, the taste difference between sun-dried and dehydrated is pronounced, though in a way I can't describe, except to say that they both are acidic and salty but the sun-dried shiroboshi taste better.

I found an advantage to starting with 2 kilograms (4.4 pounds) of plums, then separating them into ½-kilogram batches. After two years, one of my batches was definitely off—maybe moldy, though I couldn't tell what had gone wrong, just that they did not taste good. Because I'd stored the plums in small batches, whatever happened to the bad batch did not spread to the rest of my shiroboshi.

In the United States, look for ume around Easter and unripe shiro around Independence Day. If the sun hasn't come out by the time you're ready to dry your sour plums, just keep them in the fermentation crock until summer gets serious. The extra fermentation time won't hurt them, and there's so much salt that they won't spoil.

Should you grow your own shiso or find some at the market, preserve the leaves at their peak by packing them in salt. They'll keep in a cupboard indefinitely until you need them for umeboshi.

Yield: about 2 kilograms (4.4 pounds)

2 kilograms (4.4 pounds) ripe ume or unripe shiro plums
(stems removed; do not use bruised or damaged
fruit)

2 cups (more or less) vodka, in a bowl

240 grams kosher or coarse sea salt (Itoh recommends
    this 12 percent salt-to-plum ratio for beginners;
    200 or 160 grams would also work, but keep a close
    eye out for mold)

200 grams fresh or salted shiso

A fermentation crock or large, wide-mouthed glass jar,
    washed, disinfected with vodka

Weights (at least 1 kilogram of ceramic weights, clean
    stones, whatever you have)

Soak the sour plums overnight in cold water. The next day, drain
and dry the plums, then immerse them in the bowl of vodka to kill
any remaining mold spores. Remove the plums and let them dry.

In the fermentation crock, spread a layer of salt, then a layer of
plums, then a layer of shiso. Repeat this order until there are no
more sour plums. Cover the top of the plums with a muslin cloth
or food-grade plastic, and weigh the plums down with at least 1
kilogram of weight. Cover with another muslin cloth or lid, and
store in a dark, cool spot. Check on them in 3 days. A purplish
(from the shiso) plum juice should cover the plums. If it doesn't,
use your hands to massage any undissolved salt at the bottom
up to the top fruit, and increase the weight on the plums. Then
cover them with the muslin cloth again, and return in 3 weeks.
During those 3 weeks, check the plums to make sure they're submerged in brine, and check for mold. If you find some, pick it off
and cover the plums back up.

Once 3 weeks have passed and sunny days appear (when the
temperature is above 85°F, I'd say), dry the sour plums and shiso
on rattan mats (or do as I did and lay them on dehydrator trays
and baking racks that are resting on metal baking sheets). Bring
in the sour plums and shiso overnight. If the weather isn't yet
hot enough to dry the plums, leave them in their dark, cool spot
until it is.

Meanwhile, once you've removed the plums from the crock, strain the brine through a sieve, and store it in a clean jar. This is umesu (or shirosu!), and it can be refrigerated or not.

After they've spent 3 days in the sun, split the sour plums and shiso into four batches and seal each batch in a resealable plastic bag, pressing as much air out as possible or using a vacuum sealer, if you have one. Purple liquid will pool in the bags—that's a good thing, and Hachisu says it "aids in long-term preservation." Itoh recommends storing some of the sour plums dry by letting them get very dehydrated in the sun and then packing them into clean jars.

Shiroboshi and umeboshi will keep indefinitely at room temperature. Store them somewhere dark and cool for at least a year, but preferably 5.

Chicken Katsu Rolls with Umeboshi and Shiso

One of Hiroko Asakura's favorite meals with umeboshi is these chicken rolls, which she fries like chicken katsu.

The pounded cutlets shouldn't be much bigger than an iPhone. If they're larger than that, cut them into two equal-sized rectangles. Cutlets that are too large won't cook all the way to the center.

I don't have a steady supply of shiso, so I grow my own. If you too have difficulty finding shiso year-round, harvest or buy it during its summer season and pack the leaves in salt to preserve them. Store the salted shiso leaves in a plastic bag in a dark, cool place.

Hiroko serves these chicken katsu rolls whole, cut in half diagonally, then uses chopsticks to take bites of each piece. "That way, one can still see and enjoy what it looks like on the inside, the rolled red and green and white, while keeping the whole thing warm and without falling apart." The rolls can be served with rice or on thinly julienned cabbage dressed with freshly squeezed lemon juice. Oroshi ponzu and "good old tonkatsu sauce" are also great condiments, but be careful that the sauces don't overwhelm the plum/shiso flavor.

Yield: 6 servings

1 cup all-purpose flour

3 eggs, lightly beaten with a little water

2 cups panko bread crumbs

Umeboshi paste made from 10 umeboshi (see recipe)

1 bunch fresh or salted shiso

2 pounds boneless, skinless chicken breasts or thighs

Salt and pepper

1 cup neutral frying oil, such as sunflower or
safflower oil, plus more as needed

Assemble your battering ingredients—the flour, eggs, and panko—in three separate bowls, and set them aside for now. Make the umeboshi paste by chopping and smashing the plum-flesh, discarding the pits. If the shiso leaves are very large, tear them in half.

Pound the chicken thighs as thin as they can go without disintegrating; slice the chicken breasts to an inch of thickness, then pound even thinner. Lay the chicken cutlets flat on a plate. Salt and pepper each side, going light on the salt if you're using salted shiso. Lay a shiso leaf over each slice, top the shiso with a smear of umeboshi paste, and roll the chicken slices up. Dip each roll first in flour and tap off the excess, then in egg, then in panko. Transfer each roll to a plate.

Heat the oil in a large, heavy-bottomed skillet over medium heat until it shimmers.

Add the chicken in small batches, giving each roll plenty of space in the pan, and fry until golden, about 2 minutes for all sides, adjusting the heat so the rolls don't fry too quickly. I like to fry the ends by propping the rolls up against the sides of my frying pan.

Drain the chicken katsu rolls on paper towels. Add more oil as needed.

To serve, slice the chicken katsu rolls into rounds, and lay them on a bed of rice. Store leftovers in the refrigerator. They make a delicious cold snack.

V: Vanilla

Vanilla planifolia (Mexican vanilla bean) and
Vanilla tahitensis (Tahitian vanilla bean)
Orchidaceae (orchid) family
Also known as vanilla bean, real vanilla

I wanted to be a girl. It was 1995, and I was already a girl. There were, I was discovering, always new ways to be one.

This time, girlness required a practiced, steady hand to squeeze pale-yellow stripes of vanilla lotion down forearms and shins and rub them into exposed limb-skin, culminating in a businesslike hand-wringing that moisturized wrist to fingertip. The girl then slid the bottle into her backpack and sat through math class within a cupcake-scented cloud. The act and the aroma seemed to me to make the girl more visible. Especially if she pretended otherwise.

I did not want to be visible so much as not so invisible.

Most of all, I liked how the vanilla girls smelled. I had a sweet tooth. I still do. I still like vanilla cake.

Vanilla itself does not taste sweet. (Neither does lotion, usually.) It smells sweet, though, because sugar "lends" itself to vanilla, which makes it a handy flavoring for manufacturers of low-sugar diet food. The sweetness exchange is one-sided—we think of vanilla as tasting sweet, but we don't think of sugar as tasting like vanilla. According to *The Oxford Companion to Sugar and Sweets*, this is normal. "In general, taste donates while olfaction receives," perhaps because we evolved to use our noses to identify calorie-laden foods. Calories and sugar tend to correspond, teaching our olfactory organs to sense that certain odiferous fruits (like vanilla beans) are sweet whether or not they actually contain sugar. Sugarcane's sucrose can stimulate our sense of smell, but it is not itself very smelly, and so probably not the source of this association. More odiferous substances with rich flavors, like cream and fruit—which contain lactose and fructose—are better examples of how we may have come to associate sweetness with certain odors.

Another theory: we create sweet scent associations more casually, culturally, through repetition. Like many people in the United States, I identify vanilla with dessert, particularly ice cream (the first popular use of vanilla in this country) and cake. American girls of the late twentieth century were not cakes, but they frosted themselves with vanilla emollients as if they were. The association made the girls seem sweeter; if not in demeanor (I can tell you the vanilla girls did not treat me sweetly), then, thanks to the association between sweetness and femininity, in being *more girly*: a way to describe young or exaggerated femininity that's often pejorative. Like when a boss at one of my first adult jobs confessed that as a teen she'd decided to become an alcoholic instead of an anorexic because anorexics were girly, and she *was not girly*. The subtext being, I think, she didn't or couldn't perform that kind of femininity, and this cost her something at home or school or both, so she would not be weak and skinny, she would be something else, someone who wore motorcycle boots and survived her addictions, someone not girly. The comment seemed subtly poised to suggest her vulnerability while setting up what would become an unsaid but fatal difference between us—me being, by then, the girly one.

French- or custard-style vanilla ice cream differs from Thomas Jefferson's American-style (also known as Philadelphia-style) ice cream in that it adds eggs to a cream base. French vanilla ice cream is often just this custard-based vanilla ice cream. The black flecks in "vanilla bean" ice cream, especially if it is cheap, are probably not vanilla seeds, the pungent substance bakers scrape into their confections when they're getting serious about quality. More likely, they are what's left of the vanilla bean after all flavor has been extracted from it—the ground, exhausted vanilla pod.

Vanilla produces the only edible fruit in the orchid family. Native to southeastern Mexico, Central America, and northern South America, it is an epiphyte that climbs trees and dangles roots, and blooms greenish flowers once a year in the early-morning hours that wilt by noon if not fertilized. The name comes from *vainilla*, Spanish for "little scabbard," which comes from *vaina*, Spanish for "pod" or "sheath," from the Latin *vagina*. Humoral medicine considered vanilla "hot," like all New World foods. It appears in an Atzec herbal as a treatment for syphilis, and was at least as useless as the treatments of antimony and mercury that Europeans recommended for that disease, but without the dreadful side effects. Vanilla does help calm nausea. It is one known ingredient of Coca-Cola's many secret ingredients. It is the "cream" flavor in American-style cream soda and a modern addition to the mixture of herbs (or artificial flavors) that make modern American root beer, a sugary, carbonated version of Old World spring tonics traditionally made with sassafras, birch bark, and wintergreen. Aztecs, early-modern Europeans, and plenty of people today consider vanilla an aphrodisiac, perhaps an inevitability of its high price.

Vanilla orchids can thrive outside tropical conditions—in greenhouses, for example—but they will not fruit. To cultivate vanilla beans, obtain no more than a hectare or two of forested land (vanilla is not successful on large plantations) within twenty-five degrees of the equator, preferably in Mexico, Madagascar, Tahiti, Hawaii, or Uganda; hire people to guard your crop; learn how to hand-pollinate; and wait. In the wild, vanilla is pollinated, we think, by hummingbirds, *Melipona* bees, and other native Mexican insects; in a good year, it will yield just eight to ten beans. The discovery of a hand-pollination method on the island of Réunion in 1841 by Edmond Albius, a twelve-year-old enslaved child who was later freed, is the only reason cultivated vanilla is remotely productive today.

The beans mature on the vine for nine months and are harvested by hand when they are yellow-green and not yet fragrant. Then they are quickly steamed at 160°F and left to cure for four to six weeks, during which time the beans turn black with oxidation. Vanilla's famous perfume begins to develop, as do crystals of vanillin, which are much prized for their flavor. The beans are then rested on racks for another month, where they dry, producing a flexible, deep-brown pod that can be stored at room temperature indefinitely. By the time each bean is ready for market, it has been handled dozens of times, making vanilla beans one of the most labor-intensive crops in the world.

When fertilized, all varieties of vanilla orchids produce pods filled with seeds, but only two species are used commercially: *Vanilla planifolia*, the native Mexican vanilla bean, and *Vanilla tahitensis*, the Tahitian vanilla bean. Mesoamerican tribes gathered wild vanilla as a sacramental herb as early as three thousand years ago; wild vanilla continued to be valued as a medicine and spice throughout Mesoamerica's classic period. We can probably thank the Totonac people of the Misantla Valley in southeastern Mexico—who gave vanilla female characteristics and central cultural significance—for domesticating vanilla, though it is unclear when domestication happened. Some sources say between the mid-seventeenth and mid-eighteenth centuries. However, records from 1740s Papantla, the first city in the world to produce vanilla for international export, indicate the beans were still being collected from the surrounding forests—not cultivated—at that time.

We can also thank the Totonac for discovering that, as fresh, scentless vanilla beans dried in the sun, they developed an intoxicating perfume. Totonacs probably gave vanilla to Aztecs as tribute; the Aztecs called it *tlilxochitl* ("black flower") and prized it as a supporting flavor for *xocoatl*, the frothy, spicy, fragrant beverage Spanish women in the New World adopted for their own enjoyment by adulterating it with milk and sugar and serving it hot. According to Patricia Rain, author of the first vanilla cookbook and a history of vanilla, the first Europeans to send vanilla home were soldiers in what is now Cuba in the year 1502, long before vanilla was domesticated. "They thought it was perfume," Rain writes.

Cheap ethyl vanillin, first produced from the sapwood of conifers and patented in 1875 by German chemist Ferdinand Tiemann, imitates vanilla—its flavor is three and a half times stronger—but it lacks the nuance of real vanilla. Today, the pure aroma remains unreproducible in a lab. I do not know if what I smelled on the vanilla girls was real or imitation vanilla. Only 2 percent of the modern market is real vanilla, so my guess is it was fake.

But real enough for me.

I was not the sort of girl who knew where to get cool things, hot products, correctly branded clothing. Were they gifts from older sisters? Some kind of inheritance? Was there a catalogue of cool I could subscribe to? (Yes: in two years, I would find this catalogue, and it would be named dELiA's.)

My mother did not have this skill, either, and, on top of that, did not feel feminine or comfortable with teaching me the stereotypical expressions of femininity I was desperate to learn. "You got the wrong mother," she'd joke as we named the new differences between us.

That Christmas, a wicker basket of vanilla potions from

Bath & Body Works arrived with my name on it, shipped by an East Coast aunt. The contents of each clear plastic, forest-green-capped bottle smelled of baking vanilla, sweet and simple and real, like you could eat it. By then the vanilla girls had moved on to other scents: "Boudoir Pear" from Victoria's Secret. "Compost" from the GAP. "CK Dung" from Calvin Klein. I could mimic the girls without drawing their attention.

I frosted myself in private, feeling changed, *more* of something, excited—but embarrassed, too. Mimicking the ritual did not make me a vanilla girl. But it did call my attention to my own container in a way I would, from then on, pay attention in small daily moments for the rest of my life. I frosted myself without feeling on display and pretended that nothing was different, just like the vanilla girls, but for my own reasons. I was embarrassed that my mother might notice how hard I was trying. To be a girl. Applying the lotion felt like stealing scoops of vanilla frosting from the Pillsbury carton at the back of the fridge. As if this pleasure were unspeakably private, beyond the rules. But my mother wasn't blind. She could smell. Perfume gave her headaches.

She said nothing. She kept my privacy.

When I was a girl, vanilla seemed knowing. Now that I'm an adult with the memories of that girl, vanilla is a teen girl's scent I would never wear—not wanting to be taken for a teen girl, or to find out who has a taste for teen girls. One vanilla girl, Rachael, was so beautiful and breasted and *thirteen*—when I remember her now, I worry. The image vanilla brings: watching Rachael frost her tan legs with lotion after gym class, the boys and girls all riveted on her perfect, scented skin. What was high school like for Rachael? Where is she now?

I'm not being entirely fair here. Some of us just love vanilla, nothing creepy about that. Maybe vanilla is indeed, as slang would have it, a normal, boring, default, standard, lacking-of-special-features, within-the-usual-parameters, no-frills preference—a term now applied to software, finance, and sex.

To find my old vanilla I go to the mall, which, like many malls, has a Bath & Body Works store, loud and smelly and stuffed with so many products I can't focus my eyes to read the labels. Instead, I use color cues to identify the fragrance I want. Vanilla will be cream or brown or somewhere in between. Ah—I'm wrong. At Bath & Body Works, it is red and purple. Around me, people open containers, wave them under their noses, and inhale. They re-cap containers, replace them on the shelves, grab others, smell them. They hand the open containers to their companions, and the companions smell. The scents are floral, fruity, musky, sweet. They mix and mangle each other into an overpowering mush that makes me feel stupid but pleased, like we're at a cinema for noses and the only thing playing is Marvel movies.

The scent I remember, Bath & Body Works Vanilla, is extinct. The company does still make vanilla lotions, but they are not the right vanilla. They are "Warm Vanilla Sugar," a spicier vanilla that's more like eggnog. They are "Vanilla Bean Noel," a sugar-cookie scent that really does smell like sugar cookies. The "vintage" vanilla scent, sold online for nostalgic people like me, is "Wild Madagascar Vanilla," which smells not like vanilla but like cotton candy, though the bottle promises notes of "Our Exclusive Madagascar Vanilla Accord, African Pear, Wild Jasmine, White Sandalwood." The gingham pattern that used to mark B&BW's brand identity appears in thin strips around the surface edges of display tables, peeking from beneath pyramids of product inspired by—according to a myth concocted by corporate—a Midwestern gal named Kate who loved making natural body products from stuff she found on her farm, studied biology in college so she could open her own business, and now brings a heartland sensibility to her not-so-natural products. Bath & Body Works' parent company is L Brands, which also owns Victoria's Secret (where one can buy a "Bare Vanilla" iridescent fragrance lotion), whose chief executive, Les Wexner, was in 2019 embroiled with the Jeffrey Epstein scandal of sex trafficking in underage girls.

Vanilla extract has a different scent from straight vanilla beans, one that bakers call a "baking vanilla scent" as opposed to the fruitier vanilla-bean scent, or the musky ethyl-vanillin notes hidden in Shalimar, the first Oriental-class fragrance ("Oriental" is being replaced in perfume parlance with "amber," but this is still the description used by Guerlain, Shalimar's producer). One can find Shalimar at Walgreen's and Walmart, but I'm already at the mall, so I go to Macy's, where the wall behind the perfume counter is plastered with neon-orange signs that declare NOT A RE-TAIL AREA. I appear to be alone, but I am not: a woman is camouflaged behind the Calvin Klein Obsession display (also a vanilla-based fragrance). Her name is Janet. When

she asks if I need help, I jump in fright. I'm looking for perfumes with vanilla notes, I say. She says I'm not supposed
to think in notes. "Ask yourself, 'What story do I want to
tell? Does this perfume tell that story?'" She sprays Shalimar, Obsession, and Hypnotic Poison on branded pieces of
cardstock and hands them to me to smell.

I disobey Janet. I can't help it—the perfumes in my
stories aren't this complicated. I need my notes. Like: It's
hard to sense the vanilla in Shalimar, which might be the
point—it was always made to be a perfume whose elements
combined synthetically to make a scent only found in a bottle. Shalimar's top note is bergamot, though it reminds me
of peaty Scotch. The middle note is iris—root, not flower,
I assume? Also known as orris root, the violetlike scent
Helen Schlegel in *Howards End* uses to describe a tiresome
encounter with a conventional woman, as in, "She had a
face like a silkworm, and the dining-room reeks of orris-
root." Shalimar smells like the inside of my grandmother's
purse. I like it.

Hypnotic Poison's notes are jasmine, bitter almond,
and vanilla. "The bitter almond makes the vanilla more vanilla," Janet says. It smells like my friend Dominique, who
tried and failed to find me a boyfriend when we were fifteen
and kept a large tank of frilly fish burbling in her bedroom.
Where is Dominique now? I could look her up on social
media, but I don't want to. In our current moment of being
able to access anything online—including specific species
of vanilla from specific worldwide terroirs—my memory of
Dom is sweetest sealed off in our shared past, not dug up
and dissected in the present. I tell Janet that bitter almond
is an appropriate note for a scent called Hypnotic Poison,

since bitter almonds are full of cyanide. She takes a step back from me, still smiling, but says she's worried for my husband. She's going to watch the paper for my name.

I've been making my own vanilla lotion, trying to remember what that first bottle might have smelled like. It's much harder than I thought it would be. First, real vanilla fragrance is difficult to extract on short notice. I can heat vanilla beans in a carrier oil carefully for an afternoon, hovering nearby to make sure the oil doesn't overheat and ruin the fragrance. I can submerge the beans in carrier oil and store them in the dark at room temperature for at least a month, producing a truer, stronger scent. But there is no quick, easy way. Also, real vanilla beans are so expensive that my local natural-grocery store stopped carrying them in the bulk aisle, where they were easier to steal, and moved them into shrink-wrapped, brightly colored packages, two for $14. Real vanilla is, as any baker knows, a commitment of funds to quality.

Once I've prepared the lotion, the trouble compounds. The scent is, especially at first, too subtle, and the vanilla smells like fruit, not the creamy, spicy baking scent I'd hoped for. Using vanilla extract in addition to vanilla beans offers a shortcut, but the aroma is flatter and turns the lotion a dull beige, not the pale cream I want to preserve so the lotion resembles vanilla frosting. It also leaves my skin shiny with organic oils, definitely not the "nongreasy" formula that current versions of B&BW body butter promise. But I can get used to that. After five minutes, the shine fades.

Over the next two months, the vanilla lotion develops a deeper, stronger scent: identifiably, undeniably vanilla.

When activated by my body heat, the fragrance ripens into a long-lasting perfume that stays true and creamy and luscious, like walking past an ice-cream shop, but not cloying. Best of all, over time the aroma does not wither into a fake candy bouquet. It is not the lotion I remember. It is better than I remember.

As I spread vanilla cream on my skin after showering, inhaling the perfume I once coveted, a phrase from an article I read recently about moisturizers and aging replays in my mind, stuck in the moment of this ritual—how so-and-so beauty guru says it's important to apply creams to skin immediately after bathing, because "you can frost a dry cake, but it's still dry cake."

I think, mmmm, cake. Then I dress myself and get on with the day.

The first several minutes of real vanilla body cream are intoxicating. After that, unless I press my nose into my biceps, I can't smell it. I'm too close. I'm inside the vanilla now.

Whipped Vanilla Body Cream

If this description were written on the back of a Bath & Body Works bottle, it would read something like, "Luxurious vanilla whipped with decadent shea butter makes a rich, enticing body cream. Smells good enough to eat. Notes of vanilla-bean custard, yellow cake, dried fruit, and almonds."

If you can find Mexican vanilla beans (*V. planifolia*), use them. The scent will be creamier than Tahitian vanilla beans. Both types of vanilla will smell sweet.

The hardest ingredient to source for this recipe is probably time; however, depending on where you are, vanilla beans can be tough to find, too. Try Jones & Company Vanilla and Flavorings' on-line store for a global selection of exquisite vanilla beans. In ideal conditions, whipped vanilla body cream will be ready 1 month after you begin the recipe, but you can use it immediately if you want to. Unlike the vanilla lotions one buys at the mall, this cream has no preservatives. Store it in the refrigerator until you're ready to use it.

Yield: ½ cup

2 ounces sweet almond oil

½ vanilla bean, slit open along one side and halved again, each portion about 2 inches long

1 ounce unrefined shea butter

½ ounce beeswax pastilles

First, infuse the sweet almond oil with the vanilla bean. The best method is the long way: combine the vanilla bean and oil in a glass jar, cover it, and store it in a dark cupboard for at least 1 month—or 2, if you can be even more patient. If you don't have that kind of time, try this shortcut: Pour the almond oil into a small glass jar, then immerse the vanilla-bean pieces in the oil. Place the jar in a small saucepan, and carefully fill the saucepan with water so the water level is at least as high as the oil level inside the jar. Heat the jar over low heat for 2 hours, maintaining a water temperature of 115°F. Do not overheat, because that will spoil the scent. Once the oil has been infused with vanilla, dump out the water, dry the small saucepan, and pour the oil into the saucepan, reserving the vanilla beans.

Or, if you infused the vanilla oil the long way, just pour the oil into a small saucepan, reserving the vanilla beans.

Scrape the vanilla seeds into the oil and toss in the pods. Add the shea butter and beeswax. Heat over low to medium-low heat until the ingredients have melted together, stirring to hurry the beeswax along. Do not heat any hotter or longer than it takes to melt everything into one liquid.

Once everything is melted, pour the oils into a medium-sized metal bowl. Extract the vanilla beans and set them aside for now. Let the oils cool to 85°F (they'll have turned from translucent yellow to a creamy off-white), then whip them with a hand mixer on high speed for 2 minutes or so, pausing in the middle to scrape the cream back together before continuing to whip it more. When it is white and light-textured—a bit like whipped egg whites before they form peaks—the cream is done.

Spoon the body cream into a 4-ounce jar. If you like, insert a vanilla bean into the center of the lotion. Cover, and store in the refrigerator until ready to use. It can be stored at room temperature, but after a month it may start to degrade.

This cream can be used immediately, but the vanilla fragrance will increase over time.

To apply, scoop a small amount from the jar with the tips of your fingers and spread it over your skin, allowing your body heat to warm the cream and activate the fragrance. Don't worry about the vanilla-bean seeds. Once you've applied the cream, you won't even see them. Remember, compared with conventional lotions, this cream is quite concentrated. A little goes a long way.

Vanilla Bean Cake with Vanilla Buttercream

I used to sell mini-versions of this cake at a farmers' market on behalf of my friend Mika Maloney, owner of Spokane's Batch Bakeshop. I asked her if she'd share the most vanilla of her vanilla cakes, the one her customers all crave. This is it.

The instructions for how to build a layer cake are abbreviated from Mika's zine on the subject, *A Small Batch*. "This is what I tell people in my cake-building class," she says, "but with lots more hand motions and waving my arms around."

Yield: one 8-inch or 9-inch layer cake, or a taller 6-inch layer cake, assuming 2 pans for each layer cake

VANILLA BEAN SIMPLE SYRUP

    1 cup sugar
    1 cup water
    1 vanilla bean

In a small saucepan, combine the sugar and water and bring to a simmer, stirring to dissolve the sugar. Add the vanilla bean, and simmer for 5 minutes. Remove from the heat and cool overnight, leaving the vanilla bean in the syrup.

VANILLA BEAN CAKE

    4 large eggs, at room temperature
    1 tablespoon vanilla extract
    ¼ cup whole milk, at room temperature
    ¼ cup buttermilk, at room temperature
    1¾ cups low-gluten flour (cake flour is fine, or
        all-purpose in a pinch)
    1⅓ cups sugar
    2 teaspoons baking powder
    1 teaspoon salt

1 vanilla bean

8 ounces (2 sticks) unsalted butter, at room temperature

In a medium bowl, whisk together the eggs, vanilla extract, whole milk, and buttermilk, and set aside. Combine the flour, sugar, baking powder, and salt in a large bowl and whisk to combine. Split the vanilla bean and scrape its seeds onto the butter (no need to mash it in). This helps the tiny vanilla seeds distribute evenly in your batter.

Add the soft vanilla butter into the dry mixture a tablespoon or two at a time, mixing with an electric beater on medium-low speed, until the butter and dry ingredients form moist crumbs. This won't take long after you've added all the butter. Mika says you don't want to mix it completely together like a cookie dough.

Beat in all but ⅓ cup of the milk mixture. Beat on medium to medium-high speed until light and fluffy, about 2 to 4 minutes. Reduce the speed to low, and slowly beat in the remaining milk mixture. This will take about 15 seconds. It might look curdly after you add this, but don't worry.

Let the batter rest for 30 minutes. While you wait, preheat the oven to 300°F. Grease and line the cake pans or cupcake pans with parchment paper and butter: if making an 8-inch cake, fill two prepped 8-inch round cake pans two-thirds full. Drop the pan onto the counter from 8 to 10 inches to help any bubbles settle.

Bake for 30 to 50 minutes, or until the cake is deep golden and springs back when touched, and a knife or cake tester comes out clean. Remove the cakes from the oven and, as Mika generally recommends, cool them in the pans for about 10 to 15 minutes, then flip them out by inverting them onto a board, a sheet pan, or a plate, then inverting them right-side up onto a different board, sheet pan, or plate. Allow them to cool until they're at room temperature.

VANILLA BEAN BUTTERCREAM

    6 egg whites
    Hefty pinch of salt
    1 cup plus 2 tablespoons sugar
    ½ vanilla bean
    1 pound (4 sticks) unsalted butter, cubed, at room
        temperature
    1 tablespoon vanilla extract

In a medium metal bowl, whisk together the egg whites, salt, and sugar.

Heat 2 inches of water in a medium saucepan over medium heat. When it's simmering, set the bowl on top and cook the egg-white mixture for 2 or 3 minutes, whisking constantly. Whisk until the mixture is white, beginning to grow in volume, and warm to the touch, and the sugar has dissolved, leaving a smooth texture.

Remove from the heat. With an electric beater, whisk on low speed for 1 minute, then set the mixer to high speed and mix for 6 more minutes, until the bowl has cooled and the meringue is thick and smooth.

Split the vanilla bean and scrape the seeds onto the butter. Using an electric beater on low speed, slowly add the cubes of vanilla butter (make sure it's at room temperature) to the meringue. Don't worry if it looks funky for a minute! Mix on medium-high or high speed, add the vanilla extract, and beat for another minute, until the buttercream has a smooth texture. Mix on low speed for 1 or 2 minutes to beat out any air bubbles.

Keep at room temperature to use within the next 3 days, or store in an airtight container in the fridge or freezer. Before frosting the cake, bring the buttercream to room temperature and stir it with a spatula until it's smooth.

BUILDING THE CAKE

With a pastry brush, lightly paint the surface of each cake with vanilla bean simple syrup to add another layer of vanilla flavor and make sure the cake stays moist. Be careful not to make the cakes soggy.

To frost, put a dab of frosting on a cake plate and set your first layer bottom-side down onto the dollop. Scoop ½ cup of the frosting onto the middle of the first layer, and use an offset spatula to push the frosting gently out toward the layer's edge. Flip over the second cake layer so the flat bottom is now the top, and press it firmly down on top of the first (now frosted) layer. Scoop ½ cup of frosting onto that second layer, and spread the frosting with an offset spatula as you did for the first layer. Use a longer spatula to apply a thin layer of frosting all around the cake. Make sure you cover all the seams. Chill the cake in the refrigerator for 10 minutes.

Once the cake has chilled, bring it back out and scoop a cup or two of frosting on top. Use an offset spatula to press the frosting gently out from the center of the top to the sides, and then swoop down the sides. Scoop more frosting directly onto the sides if you like, or start on top again and push out and then down. The key is to push from a pile of frosting, versus trying to dab it on each spot, which would make you more likely to pull up crumbs. Be generous to start. You can always go back and scrape off frosting as you even it out.

Chill the cake again, for 2 hours, or until the frosting sets. Then bring the cake to room temperature and serve. To store, cover with a cloche and leave the cake on the counter. Eat within a week. I think it tastes best on the second day.

This cake can be frozen whole or cut into slices and frozen to eat later. Freeze the uncovered cake until the frosting is stiff, then wrap it in wax paper or plastic and seal in an airtight container. To thaw, leave the cake on the counter. It's delicious frozen, too. And with vanilla ice cream.

W: Wheat

Triticum aestivum
Poaceae (grass) family

From the ages of twenty-seven to thirty-two, I paid my rent by cutting all-purpose wheat flour with Irish butter to make pie pastry. It was an emergency-fund / self-help scheme. If I turned one kind of dough into another by the first of the month, could I spend that month at home as I wanted to spend it, working for myself—writing? Compared with the other pursuits of my life, baking pies was easy; the failure, if I failed, simple to identify and redeem. This is when the kitchen became an extension of my writing studio, where I first transformed raw fruit and pastry into something that required no explanation or argument, something almost everyone knew how to receive. I needed that then, and I need it now. Not just to have made something immediately delightful, but to be the sort of person who knows how to make something immediately delightful. Something sweet and good, then gone.

Wheat is a relatively quick cash crop with a huge market, which is as true now as it was when Civil War veterans and European immigrants rushed into Minnesota and the Dakota Territory to stake claims, raise wheat, and sell it to the mills clustered around Saint Anthony Falls. In 1850, Minneapolis milled fourteen hundred bushels of wheat. By 1870, 18.9 million bushels poured through its mill district.

Refined, bleached flour made lighter pastries and finer breads than the whole-grain stuff, which was rough, dark, and assertively wheaty, and went rancid fast. University researchers sponsored by Washburn-Crosby (progenitor of General Mills) and its competitor, Pillsbury, pitched this new product to the public by saying that bleached and bromated flour was tastier, healthier, pure.

A decade into the twenty-first century, when W and I were still in love, we discovered that if one half of a relationship is a baker and the other is a celiac, the baker must take surgical steps to contain her gluten. The celiac, for his part, must regard the baker's wheat as poison.

There are three basic parts to wheat. Technically, gluten isn't one of them. The germ, where the seed is stored, is full of nutritious oil and goes bad fast. The bran, full of fiber that protects the seed, also goes bad fast, and is sharp in a molecular way I don't really understand, except by effect: it cuts gluten bonds as they develop, making a crumbly dough. The endosperm are what's prized in white flour. Mostly starch composed of glutenin proteins and carbohydrates, these are the substances that—when developed with moisture and pressure—become the gluten bonds that give piecrust structure and flake, make the glue of wheat paste, and sicken celiacs.

I remain uneasy with the terminology of W's illness. Once he was diagnosed, in the last year of our relationship, "celiac," an appositive for "a person with celiac disease," sifted over other identifying nouns. Like how we describe people according to where they are from, or what they do. Salesman, Seattleite, Celiac.

I had a separate drawer for utensils that touched wheat, separate jars of peanut butter because I still liked to spread mine on wheat bread. W thought that if I had recently eaten wheat my kiss would make him sick. I conceded to this argument, because to do otherwise would boil down to a command, a manipulation, a plea: Kiss me. Please, kiss me.

In 1921, the marketing department of Washburn-Crosby introduced Betty Crocker, a personification of home economists and domestic scientists. Women who, rather than riot for the vote, empowered themselves by remodeling the feminine sphere with the masculine language of science. Betty had all the answers to a certain kind of question: why your bread didn't rise, why your cakes burned, why your pie dough tasted like dust. Over her first century, she aged from fiftyish to fortyish, down to thirty-five, and back up to fortyish, and was presented as a brown-haired, blue-eyed Caucasian woman except for a period in the 1990s when she looked vaguely mixed-race. She was ever new, ever improved until the beginning of the twenty-first century, when depictions of Betty largely dropped off General Mills packaging, leaving her physically erased but still present as a brand and a logo, a food goddess with no body to feed, her white name suspended in the curve of a red spoon.

My version of our story is simple. It's W's fault, even though he is sick. We cannot help how we are sick, but we can help how we live with being sick. Wheat made him sick. Every day was a series of things he could not do because he was sick, like leave the house, and things he did anyway, like drink.

Flour especially made him sick, or the idea of it. If I was making pie, he would run out of the room wailing, "You're trying to kill me!" as a joke, but then I'd find him moaning in bed, clutching the parts of him that hurt. "My guts hate me," he said. "There's flour in the air," he said. "Are you trying to kill me?"

Later, after a friend asked, "Well, were you?" I thought I'd see how it felt to say yes.

It was not him, exactly, I wanted to kill, but the way his body turned the food I made him into poison. That is what I wanted to kill.

That, and the part of him I thought took pleasure in surrendering to sickness, like it was a shelter that kept him safe.

Television for hours and hours, movies stacked up like library books, themed by actor or director, watched on fast-forward if they were bad, so he could consume the entire oeuvre without tasting the rotten parts. He had been like an encyclopedia to me when we first met, so full of knowledge I'd wanted and knowledge I hadn't known to want. Here was W with his encyclopedic brain at rest with a gluten-free cocktail, swallowing the world as it passed over the screen before us, a feast of John Ford and John Cazale and Jean-Claude Van Damme that wouldn't turn suddenly, when he thought he was being good, and kick him in the guts.

He couldn't be touched, but he would allow me to lean into him. I'm bloated, he said. My stomach hurts. He couldn't be kissed. Your breath smells like coffee, he said. Coffee made him sick, too.

I used to wake up angry, aching for sex. I used to sleep late, trying to sneak by my rage. I used to drink with him. Vodka made us feel close.

You can't say these things out loud to a sick person. It's not fair.

From the Song of Solomon: Your body is a heap of wheat encircled with lilies.

From *The White Ladies of Worcester*: If roses overgrew the wheat, we should dub them weeds, and root them out.

"I never asked you to help me," he said when I left.

///////////////

Wheat dust, a by-product of the flour-milling process, is more explosive than gunpowder.

On the day it blew up, Washburn A Mill was the largest mill in the United States, producing one-third of all of Minneapolis's flour output and employing over two hundred people. A jury investigation by the coroner of Hennepin County into the events of May 2, 1878, found no evidence that "the mill was being run in an unusual manner in any respect" and determined that "it is not possible to fix upon anyone blame for special neglect or carelessness on that occasion." They accused air purifiers—which seemed to have done the opposite of what their name implies—of causing "a needless amount of flour-dust to settle throughout the mills, stored ready for an explosion."

It was no one's fault, then, that an hour after the night shift started, two unfed milling stones shot a spark into a pile of middlings, where it smoldered. It was no one's fault that an unlucky draft of flour-soaked air happened across the blaze and exploded, which threw up dust nearby, which exploded, igniting a chain reaction of fire and flesh and stone and wheat that did not stop until it had consumed the entire building and three neighboring mills, killing eighteen men, including the entire night shift of the Washburn A.

It is true that the idea of wheat became a poison that W and I both handled. It is true that, at any given time, I kept ten pounds of wheat flour in our house. It is true that sometimes, especially at the beginning, the immediate *after* of his diagnosis, I would forget he couldn't eat wheat and find some incidental way to add it to our meal: gluten-free fish tacos battered with beer, cornbread made with contaminated cornmeal. It is true that sometimes I'd get so frustrated by the job of adapting gluten-full recipes that I'd sink to the floor and weep. It is true that it took me time to accept that we had to worry about contaminated corn, that a sack of yellow meal, sunny and humble and supposedly our alternative, would also poison him if it didn't have the magic symbol on the back: a stalk of wheat jailed in a crossed-out circle. It is true that he was the axis our meals turned around.

Also true: I turned the wheel of our meals. I liked the way my hands looked, steady and capable at the controls, secure in the fiction that he couldn't pilot this route himself. Did he cook for himself when I met him? Did he cook when we were together? All I can remember are his gluten-free waffles, torn from the bag and toasted before work.

When the new Washburn Mill opened in 1880, it was equipped with filters that sucked away the fine dust chuffed up by millstones every day and pumped it into the Mississippi River. For a while, that meant bran and germ, bugs, rat shit, pebbles, who knows what else, until a secondary market in pig feed made from germ and bran removed those substances from the waste stream. But, still, it's said that the flour dust formed islands of dough that floated downstream, where they collected in eddies and rotted. Catfish gorged themselves. You've never seen catfish so big.

"Water + wheat flour = dough" was true on skin, too. A sweaty man who spent his workday hauling fifty-pound flour bags would have to comb his arms to "get the goo balls off" at the end of the day. If he worked in the part of the factory that mixed flour by-product with blood meal to make animal feed, the dough that covered his arms would be bloody.

The same equation was at work in the lungs of mill-workers, who would breathe in wheat dust and cough out dough. They called this "miller's lung."

This is incorrect. They did not cough dough, they coughed blood. Miller's lung is cured by the kind of government regulation that produces decent working conditions. As recently as 2007, millworkers in Nigeria and Iraq reported symptoms of the disease.

Perhaps this is getting too gruesome. Remember that mill jobs were good jobs, with steady and respectable paychecks; that wheat milling was, as *Minnesota History* describes it, "a basic industry, meeting a fundamental human need." Remember that what's interesting here is the contagiousness of the conflagration, the contagiousness of flour, the way just being physically near it implicates the skin, the lungs, the body, but that wheat's harm is so fantastic or so gradual, and its benefits are so widespread and believed in, that it remains a welcome neighbor, a family staple.

It would be easy but wrong to call wheat simply poison, simply a contagion of no nutritional value. Like we're members of the gluten-free church, those people who preach, "There's hardly an organ that is *not* affected by wheat in some potentially damaging way" (William Davis, *Wheat Belly*), and sell their diagnoses by promising, "Elimination of this food . . . will make you sleeker, smarter, faster, and happier." Purity of this kind is impossible. It is an idea, not a cure.

Can we also say that part of the weight of being ill is protecting the people you love from your suffering? Within relationships, there would be sickness in containing your suffering too well, and sickness in not containing it at all.

Can we say, too, that when we are ill it can feel *good*—especially when medical remedies aren't satisfying—to give some of our suffering to others, to make it just a little bit their fault? Suffering would speak, like a virus, in a language of silent codes that replicates itself through how a suffering person makes the people he or she loves suffer. The caretaker's sympathetic response would become, over time, a source of chronic pain. "Makes" implies action—the sick person spreading their suffering around. I don't mean that. I don't mean W wanted me to suffer.

If it is funny to tell a story that says I tried to poison the man I loved, it is because this story lends a pleasing shape to how we fell apart: something rectangular and papery, leaking from the corners but basically sound, as blameless as a bag of flour.

No, I did not try to poison him. But I would not stop baking.

I could feel the question coming: would I stop, for his sake? I could feel him hold himself back from asking, the plea expanding in our air. The answer was a line drawn between us. I would stop, but I wouldn't forgive him for asking. Or I wouldn't stop, and he would blame me for baking him sick. Either way, a conflagration of resentment.

After I was sure he had moved out of our house, after our room was stripped of bed and bookshelves, after he took his movies and music, his guitar and voice and hands, it was my turn to box up my possessions. When I returned, I saw he'd hardly touched the kitchen.

I'd forgotten: everything there—the pots, pans, silverware, food—was mine. I took a bag of flour down from the cupboard, reached into its dry white innards, withdrew a fistful of the stuff that chased him out of this room. I flung it on the counters, on the stove, in the sink. I smeared it on the knives and bowls and plates. I dumped it on the floor, in each groove of tile and seam of air, everywhere. He was gone, and there was flour everywhere.

Whole Wheat Pie Pastry

Yield: pastry for one double-crust pie

To make flaky piecrust with whole-wheat flour, appreciate that whole-wheat flour is like coffee: it tastes enormously better when freshly ground. So, first, buy a wheat grinder and a bag of soft spring wheat. Second, grind the wheat. Two and a half cups of ground flour will do. Add 1 tablespoon sugar and 1 teaspoon salt, and mix. Third, make a mixture of ½ cup cold water and one egg yolk (save the egg white to glaze the top of the pie before baking), beat to combine, and freeze while you prepare the rest of the dough. The proteins in egg yolk will help this crumbly dough roll out and add tenderness and richness, though the dough won't be short on richness, because, fourth, cut 1 cup chilled unsalted butter into ½-tablespoon-sized chunks and rub them into the flour until the mixture resembles chunky cornmeal, with some pea- and almond-sized lumps of butter-flour, the flour and butter not combined—not exactly—more like *containing* each other. Fifth, pour the icy yolk-water over the dough a little at a time, and toss by hand to distribute the moisture evenly. When the dough *just* comes together, stop adding liquid (there will be some left over), stop tossing. Sixth, gather and quickly press the dough into two thick discs, cover them, and refrigerate for 30 minutes. Seventh, roll immediately, after dusting your rolling surfaces and pin with all-purpose flour. Pie dough left in the fridge overnight will become hard and stubborn and impossible to work with. Finally, eighth, adjust your expectations. This dough won't roll out as easily as dough made with all-purpose flour, and when it's baked, its flakes will crumble rather than sheer. It will taste nutty, cookielike, especially if the flour is fresh.

Wheat Paste

Stronger than household glue, wheat paste is a favorite of guerrilla artists and primary-school art classrooms. For extra strength, add a spoonful of sugar, and then beware: once this stuff is smeared on, it's very difficult to scrub off.

Yield: a little less than 2 cups

4 cups water, divided
3 tablespoons all-purpose flour
1 tablespoon sugar

Pour 3 cups of cold water into a medium bowl, then slowly pour the flour into the water, whisking as you go, to make a slurry. In a medium saucepan, bring the remaining cup of water to a boil. Pour the cold mixture into the boiling water slowly, whisking as you go, then return it to a boil. Continue to whisk. After a few minutes, when the mixture has thickened, remove it from the heat, whisk in the sugar until it has dissolved, and pour the wheat paste into a disposable container. Allow it to cool before you use it. To spread wheat paste over a surface, use a paintbrush.

If a portion of the paste is left after you're done, store it in the disposable container with an airtight lid and refrigerate. Wash all pots, utensils, and brushes immediately after you're done using them.

X: Xylitol

A sugar alcohol created by plants and animals,
including humans
Colorless or white crystalline solid
Also known as birch sugar, xylit, xylite

"Our bodies are made of this," my stepdaughter says as she cradles my new crystal, a smoky quartz she cooed over in a way that made me feel I'd won something. "See these?" Jane holds the quartz so I can see the internal fractures. They look like knife cuts in Jell-O and spark when they catch light. "Rainbows," she says, the way a forager might name mushrooms. "Here and here."

I don't know how this can be true, our bodies being made of quartz, except that the astronomer Carl Sagan said we're all made of "star stuff," and I enjoy believing him, so why not?

Is this any weirder than insisting, with a mouth full of carrots, that we are what we eat?

I bought the quartz because Jane is into rocks, all their energies and properties. She talks about them like they're alive, which has us worried, but no more worried than if she'd started going to church. Something's wrong, something hurts, something's too beautiful for words, something's changing, and rocks are a way to name things.

It's appealing, how she describes the energetic properties of quartz and celestite and apatite. Like anyone could use them to purify and channel inner weather, or even predict it—the way keeping a remedy on hand predicts the problem to be remedied. How Sam carries Tums in his back pocket because he knows that after a Friday-night feast at our favorite restaurant, I'll need them.

When Jane's working at the rock shop, I visit and ask for something that will help me write this chapter. She gives me turquoise for communication and truth, and aragonite for deep grounding. "Put these by your computer," she says, "or in your bra." When I ask her what she thinks about the difference between drugs and medicine, she says, "I'd call the white powder in capsules I pick up at the pharmacy 'drugs,' and weed 'medicine.'" Her parents are more uneasy about the weed than about the crystals, but she is an adult, and these are her choices, her medicines.

Another time, after a highly rated tailor I'd found on Google asked—while I was in his dressing room, putting the formal dress he'd just marked for hemming on a hanger

and scrambling into my jeans—if I was married, if my husband was tall and strong, and said that if it didn't work out
he'd take me out, no problem, blue eyes, something else I
refused to hear; after freezing like a clichéd headlit deer; after abandoning my dress and money, nodding and smiling
until I could leave; after weeping with rage on the sidewalk,
feeling so stupid, why didn't I grab the dress and storm out;
after deciding, fuck it, I'm not sorry, and calling Sam to
get my dress and money and yell at that horrible man, a
task Sam enjoyed very much—after all that, to feel better, I
visited Jane at work for a crystal. "Citrine, that's what you
need," she said, handing me a glassy yellow point and telling me to put it, too, in my bra. "It's personal power, self-
confidence, strength, and abundance," she said. "Fuck that
guy," she said. "He doesn't get to put his gross energy all
over your nice dress."

I'm not sure for how long, maybe three years, my friend
X said she had Morgellons disease. Tired, achy, nauseated
with no explanation, haunted-looking, with wires and filaments extruding from her skin. I couldn't see the wires
and filaments, but I decided that wasn't important. Clearly,
she was sick. Clearly, something was wrong. The important
thing was to believe her when she said she was in pain.

When she stayed with me, X would set up a small altar near where she slept, anchored by a crystal Jane would
have loved—but this was before I knew Jane. Among other
things, X used the crystal to purify tarot cards. I didn't ask
often, but when I did ask, X gave me readings. My questions were always, at their root, the same: *Who's going to
love me? Will my art be worthy? How will I make a living?*

The answers generally confirmed what I suspected, and changed according to whether that day I most feared loneliness, failure, or poverty. "The cards aren't psychic," X said. "They just tell you what you're ready to hear." *As in, they tell me what I* want *to hear?* I thought but did not say. I didn't believe in tarot, but I liked the way X helped me feel I was getting somewhere. Even while so much felt out of control, my life was within prediction.

I liked that when it came to the cards, X didn't care if I truly believed.

In the early fall of 2016, the poet CAConrad, in a series of what they called somatic poetry rituals, taped a murdered lover's favorite crystal to their third eye, swallowed a smooth "worker-crystal" before going to bed, and let the stone move through their body overnight, letting the lover's essence, their memories of him, everything they loved, flood their "bones, [their] tissue and blood," pump "his library through [their] heart and thoughts." The next day, Conrad shit the crystal out, sterilized it, and repeated the ritual. "Almost immediately my body calmed," they write, "every cell dropped their heads back and sighed. The stress of loving a man murdered without justice lifted each day of the ritual toward peace." This blog post was passed around my social-media feeds by fellow fans of Conrad in 2016, and I thought at the time, as I do now, *This is nuts.*

But I admire it. The clarity of Conrad's image, of a crystal realigning cells like a magnet through iron filings, putting one's house in order, is satisfying. I couldn't swallow a crystal or shit it out, not even once. That's not my kind of medicine.

///////////////////////

My mother has always used diet to treat her health problems, so much so that diet, for me, connotes a way to ease
headaches and gastrointestinal distress even more than it
means "Let's get skinny." As she ages, she gets more serious
about her diet books, and they get more serious about her.
Alkalize or Die is a favorite. *How Not to Die* is another. She
keeps them on a bookshelf that is the first thing I see when
I leave my parents' guest room in the morning, shelved
next to books my brother and I abandoned over the years
and a couple books I wrote myself.

On the one hand, I admire the directness and revo-
lutionary zeal of the titles. Besides "How to Sit Still" and
"How to Go to Sleep When Sleep Reminds You of Dying,"
isn't "How Not to Die" the primary lesson we give our children when we nag them about bike helmets and looking
both ways and no candy from strangers?

I don't know if Mom's legumes and cruciferous greens
and turmeric shakes are going to help her live longer than
she would have. I guess it's not about living longer. "She's
the healthiest sick person I know," my brother says. "I'm
feeling better every day," Mom says.

When Sam's in his seventh day of his first attack of gout,
still walking with a cane, I call my mother, whose health
advice I trust above all others', even if I don't follow or always believe it, and ask what he should do.

"Stay away from meat," she says. "It acidifies the body."

"*Got it*," I say, the way I say "Peace be with you" in
church and mumble "Namaste" after yoga.

324 THE BOOK OF DIFFICULT FRUIT

Doesn't hurt to say the words, even if "acidifying the body" sounds like an invented set of reasons to avoid the meat Sam should avoid anyway. Like putting crystals in my bra for protection.

Speaking of crystals, Sam had a urate crystal in his toe right then, built by genes and rich eating. Maybe Jane was right about quartz, about our being "made of this."

Never once have I heard my father complain about hurting. Any kind of hurt. In any way. The face he wore during his father's funeral is the same face he wore during my wedding ceremony—calm and composed, with wild emotion seeping out at the edges. Not tears, never tears, but the extra shine of his eyes.

My father's medicines are different from my mother's. As a child, when I had my "polar ice melt is going to drown us in our sleep" dream, he did not say everything would be okay. He said, "We live on a hill, we'll see the water coming." When something breaks, he fixes it. Arrives to each visit with his toolbox to tackle the list of projects I've made for him. Sometimes he asks for "blue beauties" (acetaminophen), but he says it like they're a treat, like *Can I have some?* I make sure there's beer in the fridge and a sandwich at noon and we are good to go.

Sam is also stoic when it comes to pain, but with that first crystal of gout the intensity of his pain was obvious—he could barely walk. It's impossible to describe how bad this hurts, he said, and I said I understood.

But, still, while walking around the bed to get a glass of

water at 3:00 a.m., I forgot to move carefully and bumped his toe. He woke immediately with a huge suck of breath and clutched his foot. I apologized—I felt awful. I thought, *Could the pain possibly be that bad?*

Four gout-free years later, if I even brush Sam's foot with a blanket, he flinches. Witnessing how he remembers his pain is more empathy-wringing than witnessing his actual pain.

"You're being histrionic, but I'll hold your hand," Sam's friend says. I just sank my pinkie into a mandoline blade while slicing lemons for our salmon dinner. This moment—where I'm bleeding and confused by the blood and feeling myself start to faint, where I'm told I'm being histrionic by someone who, previous to this crisis, has been trying to make a good impression—all I can think is, *What a dick!* with a degree of affection that surprises me. I know my pain is real. I don't really care what this man thinks about how I express it. He offers me comfort I haven't asked for while telling me I shouldn't need it, which shows more about how he handles the pain of others than about how I should handle my own pain. *And I'm about to faint!*

I lie down on the couch and hold out my good hand. Histrionically, I hope. I make him hold it longer than he wants to. If this is his version of being sweet, it is my intention to accept that sweetness until it sours, then have a nice dinner and talk of it no more.

At some point, X stopped saying she had Morgellons and said the problem had always been Lyme disease. This seemed like an improvement. No more wires, same fatigue, but her skin looked better. Less sallow. Here was a diagnosis that many doctors and most people believed was real. She had always wrapped my couch in a sheet brought from home before settling in for a hangout; this kept happening, and that was fine. When she didn't have a sheet, I'd get her one of mine. I don't even know what she was protecting herself from. I didn't ask. There were things we agreed on, and things we could carefully probe from the other. For the rest, we suspended disbelief.

Can we agree it is painful to see other people in pain?

That how we perceive the pain of *others* is related to how we handle the pain their pain causes *us*?

That how I deal with the pain of others will probably, while I'm attempting to provide comfort, cause at least a little pain?

And I might, after a long time trying very hard to cope with the pain of others, get sick of it—sick *with* it, sick of *them*? That I might make others sick with/of me?

Isn't this normal? Isn't this true?

Isn't how we deal with other people's pain and how they deal with ours the way we create and confirm our trust in those who care for us and those whom we care for?

Isn't this the seam by which a relationship might break open? Or apart?

The years of this book dig into my muscles and joints when I sit down to my laptop or look down to read. The only thing that cures the pain is not writing. The anxiety in my stomach and guts, the tension in my arms and neck and hips, my jaw, clenched all night—when I'm really bad, I feel trapped in my body, caught in a net of ache. Imagine that, first, as a fisherman's net, bossy and binding. Then imagine it as a hammock, something I've carefully arranged to hold me in a familiar shape. Remember to breathe, the women say. Remember to eat, the men say.

"If you ayl anything," wrote Nicholas Culpeper in *A Physical Directory* in 1649, "everyone you meet, whether man or woman, will prescribe you a medicine for it."

Like a lot of people who find themselves trying to solve other people's problems, when it comes to my own I'm often caught without a remedy. As Sam digs through his pockets for antacids or boils water for my tea or rubs my shoulders, he never says, "This wouldn't happen if you . . ." or "Know what you need to do?" My mother is sneaky with bossing me about my health, disguising her concern as an exercise lesson or a recipe unless she's too overwhelmed with dread to contain herself. My father teaches me how to grill salmon and braise pork. He hangs shelves so I can line my studio with flowering plants. Jane gives me yoga lessons and translates my moods into crystals; I ask her for help not just for my own healing but because treating her as my teacher is a way to build our relationship, my new-mother to her grown-child. Sometimes I wonder if I'm compelling her to be my cheerleader, or connecting too much through pain instead of joy. "Remember your intention," my therapist says—at which I think, *My intention is to be family.*

What if love is, on some level, an exchange of medicines? The space we give and the belief we extend as the people we love wrestle with their medicines, and we wrestle with our own?

"The light in me honors the light in you," my YMCA

yoga teacher says, bowing from a semicircle of crystals, to which I think, *Jesus H. Christ,* but say, "Namaste."

It's been several years now since Jane worked at the rock shop. She no longer talks about crystals like they're alive, but I still have her citrine and turquoise and aragonite. They're next to my computer and in my bra as I write, doing exactly what she said they would do.

Xylitol Sinus Wash

Xylitol is an alcohol derived from birchwood that's refined into a crystalline form, like granulated sugar. Crystals—that's the connection.

From Ecclesiasticus: "Was not the water made sweet with wood, that the virtue thereof might be known?"

Xylitol isn't absorbed by the human body, so it's used as a sugar replacer for diabetics and dieters. It is also a natural antibacterial agent that a 2017 study published in the *American Journal of Otolaryngology* described as decreasing "the salt concentration of human airway surface liquid that contains many antimicrobial substances, which can contribute to the improvement of the innate immune system, and thereby prevent airway infections." Xylitol helps neutralize cavity-causing bacteria when chewed in gum after meals; inhibits ear-infection-inducing bacteria when fed to small children as a preventive measure; and tricks dogs' endocrine systems into releasing a flood of insulin, which causes catastrophic low blood sugar that can ruin their livers and kill them. Do use xylitol if you are a human with diabetes. Do chew sugar-free gum laced with xylitol for fresh breath and fewer cavities. Do not leave xylitol-laced gum in your purse on the floor if you are a dog owner. Xylitol is poisonous to cats, too, but cats won't burgle your purse for gum. Add xylitol to a neti pot to treat chronic rhinitis (runny nose and sneezing), or buy a xylitol spray from your nearest natural-remedy aisle. Xylitol is made much like granulated sugar, by crushing birchwood instead of sugarcane, extracting the juice, and processing that juice into fine crystals.

To re-create the remedy studied in the *American Journal of Otolaryngology*, dissolve a scant tablespoon (12 grams) of xylitol in 240 milliliters (slightly more than 1 cup) of warm water in a neti pot or nasal irrigation bottle and irrigate your sinuses once a day. The study used xylitol from ACROS Organics, which is available for purchase online. A natural-food store will carry medical-grade xylitol in packets to be used for nasal irrigation.

Xylitol Peppermints

When boiled to a hard crack and poured onto a baking sheet, xylitol will slowly form crystals, starting at nine or ten pinpricks in a clear pool of syrup and spreading outward overnight in a feathery pattern, like ice forming on a glass table. Once the entire pool of xylitol has crystallized, break it into shards and eat the shards like after-dinner mints. Or, as the circular crystals form, pry them out, set them aside to dry, and return in five or six hours to do the same for all the new crystals that formed in the remaining syrup. The "natural cooling effect" touted by xylitol marketers is not an empty promise; the mint in these candies is a physical sensation as much as it is a flavor.

Yield: about 1½ cups

1 cup xylitol
½ cup water
2 teaspoons peppermint extract

Combine the xylitol, water, and peppermint extract in a heavy-bottomed pan. Heat over high heat, stirring to combine, and boil madly until a candy thermometer reads between 310° and 320°F. Pour the hot candy onto a metal baking sheet or a silicone mat. Let it cool overnight. At first, and for a long while, the candy will be clear and syrupy, but over a 24-hour period, crystals will develop. After a day, the entire pool will be opaque and hard. Scrape it up with a metal spatula, breaking the candy into shards. Store at room temperature. Keeps indefinitely.

Y: Yuzu

Citrus junos
Rutaceae (citrus or rue) family

Yuzu, a yellow-green Japanese citrus fruit about the size and shape of a clementine, cannot be found in the same season-less grocery stack as lemons and limes, the fruits we reach for year-round to add a quick hit of acidity to our cooking. Compared to ume, yuzu's uses are minor, though delicious—in Japanese cuisine, yuzu can be found in essential condiments like ponzu sauce and yuzukoshō, as well as in candies and marmalade. But when heated, cut, or pressed, yuzu *peel* ranks with the best citrus, smelling of lemons and pine needles and sugar with a revitalizing acidity I'd almost choose over the scent of morning coffee. Yuzu fragrance is sweeter than the scent of conventional lemons, but not quite as sweet as a Meyer lemon. Like vanilla beans, yuzu peel and juice are not sweet to the tongue, but they're adept at giving that impression to our olfactory senses.

All members of the citrus family can be traced to China, where three original parents grew at least five thousand years ago: the inedible but fragrant citron, the thick-skinned pomelo, and the appetizing mandarin orange. Over millennia, the hybrids produced by this triad have created a sharp, sometimes sweet, always acidic profusion of fruits that in English are called hesperidia: berries with tough, leathery rinds named for the ideal garden in the far west of Greek myth where the daughters of Hesperus, the Evening Star, grew and guarded an orchard of golden apples (probably not apples as we know them, but citron or quince, depending on which scholar you talk to). The citrus family is a fertile bunch, mixing genetic material freely to produce a host of mutants and chimeras. Within each citrus seed resides the prospect of a completely different tree. To grow a true copy of any given cultivar, cut a scion from the parent, splice it into the rootstock of a citrus variety that's resistant to illness and frost, and bind the graft until the two grow together.

Not all citrus cultivars produce edible fruit, but a fruit's degree of edibility does not affect its degree of being fruit. A citron, for example, is mostly fragrant peel and thick pith, with dry vesicles at the center, but it is still a fruit. Its flesh is usually ignored in favor of the culinary and religious uses of the peel, which can be zested or cut or left alone so the unmolested fruit can perfume the entire room. Yuzu shares these qualities with citrons, but, thanks to its sweeter, fleshier parent, the mandarin, it contains more flesh and juice.

But not that much. Like the citron, yuzu's best qualities are skin deep.

Between citrus zest and citrus fruit is a tissue called albedo, often referred to as pith but technically known as the mesocarp. Albedo, from the Latin *albus* for "whiteness," is also an astronomical term for the light reflected by a celestial surface. Measuring this light allows astronomers to describe the properties of moons, planets, and asteroids, but not stars, which do not reflect light but, rather, emit the light necessary for an albedo to exist at all.

Bruno Munari riffs on this linguistic overlap (the same in his native Italian) with the cover of *Good Design,* a minutely described, deadpan investigation of citrus and peapods, perfect objects "in which the absolute coherence of form, use, [and] consumption is found." An orange fills the otherwise black cover, half shaded like a moon about to duck behind Earth's shadow, albedo (pith-ily speaking) tucked out of sight beneath its cratered skin. All citrus have these moony pocks and pits, which contain aromatic essential oils that have been used to enhance perfume, food, beauty, and cleaning products the world over for thousands of years.

Yuzu trees are more cold-hardy than their orange relatives, teasing temperate-climate gardeners into hoping for a real backyard citrus harvest. They are slow-growing trees, needing at least ten years before producing their first fruits, which appear in November instead of in the dead of winter like most citrus. They can be planted in pots and wheeled indoors during the cold season, as lemon farmers in northern Italy did for centuries before the technology to transport lemons from countries farther south decimated their business in the early twentieth century. Patient people with green thumbs who live in hot climates with cold winters can still pull this trick, though with yuzu it is possible to plant the tree in the ground if you live in a place with low humidity where temperatures rarely dip below freezing, like the maritime Pacific Northwest. Still, I wouldn't bet on reaping a generous harvest in Portland or Olympia or Vancouver, B.C. Be advised that yuzu trees have thorns. While picking fruit, beware.

Yuzu peels can be transmuted into marmalade, like any citrus peel, a fussy process of soaking and boiling and candying that rewards with a soft yellow preserve. Folk remedies prescribe stirring yuzu marmalade into hot water to make a soothing elixir for colds, presumably because yuzu contains vitamin C—though perhaps not, since vitamin C diminishes with processing and long storage. My favorite preparations of marmalade concentrate zest into a sweet, bracing, slightly bitter confection. The trick is to make sure the peels are completely cooked all the way through before adding sugar, which will toughen a partially cooked peel over time, and to use the albedo—a major source of commercial pectin—to thicken the marmalade.

Like citrus albedo, citrus seeds contain pectin, but to make a pleasant marmalade you must remove the seeds. Any seeds a less-than-fastidious marmalade maker misses will be a tough surprise on future toast. They are, as Bruno Munari writes, "a little gift that the production offers . . . No one, or very few, take it upon themselves to plant orange seeds, but . . . the idea of being able to do so, frees the consumer from any frustration complex and establishes a relationship of reciprocal trust."

My father's younger sister, one of two children who've lived outside our family for almost forty years, has a citrus tree in her backyard that bears sweet orange fruit with a snug peel. Jen doesn't know what kind of oranges they are. We don't really need to know.

They are green when I meet her for the first time in a house she bought the previous year, a bungalow with a tidy kitchen that looks onto a living room with a large fireplace she's filled with candles. Down the hall, her master suite is carefully and exuberantly jungle-themed, with so many cat prints on the bed, rugs, walls, and fixtures that stepping over the threshold feels a little psychedelic, like I'm inside the outside of a leopard. Jen keeps a clean house, she says. And it's true, she does. Their parents made sure they knew how, she says. "They put us in charge of cleaning and laundry, but my mom was the cook." A good cook, she says. "They had us working. And they were always working."

It is early November in Phoenix, Arizona, sunny and eighty degrees. My aunt's backyard pool gleams, as immaculate as a baptismal font, ringed with chairs she's covered in her own handmade rugs. While we talk in this oasis, under her orange tree, her roommate stays in his room, giving us space. He shouts hello when I arrive; he'll shout goodbye when I leave. Jen smokes a Marlboro. I try to resist the urge to bum. She asks if I'm nervous. On the taxi ride over, I was scared enough to text a friend my aunt's address and the time she could expect me back at the hotel, as if I were taking safety precautions before a hike into the desert. Jen is nervous, too, but she's okay. "How old are you?" she asks. Thirty-six in ten days, I say. "I can't figure out who you look like," she says. "Your mom and dad both, I guess."

Just as our knowledge of Jennifer froze around 1980, her knowledge of us did, too. She still thinks my grandmother Loretta's first husband died in World War II. She doesn't know my grandfather and uncle didn't speak for years. "Does your mom still have that long hair?" she asks. Not since 1982, I say. "Your dad, is he healthy? Is he losing his hair? Is it still curly?" Her brothers were serious boys, she says. "That's not a bad thing," she says. "Does your dad support you? I bet he does."

The night my grandparents kicked her out, she says, she had tickets to Fleetwood Mac in Cedar Falls. The Tusk Tour. Her parents forbade her to go. She went anyway. While Christie sang "Say You Love Me" and Stevie sang "Landslide" and everyone sang "Go Your Own Way," my grandparents packed up Jen's belongings. When she came home, all her stuff was on the lawn. She was sixteen.

After that, she rented an apartment, lived with friends, worked two jobs. I don't know if she stayed in school. Nor do I know whom she turned to when she was lonely or low on cash. I do know that when my grandmother was out, Jen would come by the house to visit my grandfather. "Me and my dad were tight," she says with more warmth than I'd be capable of if I were in her shoes. This fills me with surprised joy—that, despite everything, she still adores him.

"I don't know who wrote that obituary," Jen says, tearing up, referring to my grandfather's, which named only three of his five children, "but it sucked."

"I wrote it," I say. She looks at me. She's gone very still.

"I'm sorry," I say. I reach for my aunt's hand. She lets me take it. She says nothing. So hurt, but not angry with me. She doesn't seem bitter. She doesn't even seem sour.

She tells me I'm a good kid.

This January, Jen's citrus arrives by post and sits by my front door in a cardboard priority-mail box for three days until I recover enough from the flu to store the fruit properly. They are tennis-ball-sized and bright orange, leaves curling from their tops, so fresh that their trip up north and delay in my hallway didn't harm them at all. Some say citrus desiccates in the refrigerator, but my experience has been the opposite, so into the fridge they go. They smell like orange candy when I peel them. They spatter juice on my keyboard when I tear them into sections.

To preserve the oranges, I use my recipe for yuzu marmalade—these citrus are related, after all, and, when chopped and boiled and reduced in sugar, will behave pretty much the same. The first step for either cultivar is to peel the zest from the albedo, then trim away any albedo that still clings to the zest. The second step is to remove the albedo from the fruit and chop the fruit roughly, then chop the albedo roughly and set it aside in a cheesecloth bag. The third step is to boil all these citrus parts in water until the peels are soft and translucent and the albedo has released its pectin. Orange vesicles will loosen and drift, just as yuzu's do, and their color dims a notch—dark yellow if this marmalade were yuzu, medium orange for Jen's fruit. To soften, the peels will need to boil longer than feels reasonable. While they simmer, the seeds I missed will wander to the top of the pot, where I fish them out and flick them into the trash. Then I remove the bag of albedo, add sugar, boil hard until I spy small translucent lumps of pectin clinging to the back of my spoon, and pot the marmalade in sterilized glass jars as usual.

As the lids on Jen's marmalade pop behind me in the kitchen—calm, small noises that tell me my efforts to save them from spoiling have worked—I write at the kitchen table. I'm still dissatisfied with the answers for why we cut this branch from our family tree. I'd assumed Jen of all people would tell me, but not even she can say why. "After all these years, me and Julie still don't understand," she said, "especially once we had kids."

Perhaps my experience of searching for a reason is not so different from my aunts'. The same people who loved and raised my father, who loved and nurtured me, who lifted our family into the middle class, treated their daughters like bad fruit they could throw away. Or—as I have also heard over years of pestering my elders—my grandparents tried to heal the relationship, but the break spread. We lived with that loss, unable to speak of it, while our family grew lush around the scars.

My conversations with Jennifer did not unearth the rotten secrets I had feared since childhood. Instead, I find myself trying to accept what I already know: in the years right before I was born, my family faced the edge of our ability to care for each other, to take care of each other, and turned that edge into a wall.

People aren't fruit, and families aren't trees. Of course, I know that. This real person, my aunt, sent me real fruit from her hard-won present to preserve in sugar and glass and take with me into our future. To my family—her family—table. If I have any rights in this relationship, it isn't to forgive the break, nor to heal it. It *is* my right to be her relation. Her long-lost, now found niece.

I put the marmalade in a cupboard to thicken for a few weeks, as all preserves with a good amount of pectin will do. Then I wrap four jars in padding and nestle them into a box. The marmalade arrives at Jen's doorstep two days later. Next year, she says, she'll send me more fruit. From now on, she says, she'll send fruit every year.

Yuzu Marmalade

I wanted a recipe that captured the intense scent and flavor of yuzu's rind and could be adapted to work with other members of the citrus family. This one is painstaking—it asks you to separate the flavedo (zest) from the albedo (pith), the albedo from the fruit, then use all the parts to make the marmalade—but it concentrates citrus flavor into a dense preserve. When you peel the zest from the pith, yuzu may leak oil onto the cutting board and knife. Preserve that oil by chopping the yuzu pith on the oily cutting board, where it will soak up every last drop of flavor and scent. Wearing latex gloves while doing this protects your hands from acid and helps you grip the fruit, but if you choose this protective route, you won't have that incredible oil seep into your skin. I rubbed it from the board onto my wrists like perfume.

If you can't find yuzu, keep in mind that this recipe works well with all citrus that have a fairly thin peel, and works best with citrus that seem sweet. Meyer lemons will work, but conventional lemons will produce a sour preserve (of which I am a fan, but not everyone agrees with me). In terms of peel thickness, think Cara Cara oranges or medium-skinned lemons, not pomelos or citrons. Grapefruits would work. Or blood oranges. Over time, the bittersweet edge on this marmalade will soften into simple sweetness.

Yield: about 40 ounces

3 pounds yuzu or other fragrant citrus (preferably organic)
3 cups water
4 cups sugar

Place a small plate in the freezer; you'll use this later to test the set of the marmalade. Then prepare a deep canning pot with boiling water—enough to cover six 8-ounce jars. I add a little white vinegar, to mitigate the powdery white residue my hard water leaves on my jars after I boil them. Sterilize the jars by keeping them immersed in the boiling water for 10 minutes, then set them on a clean towel to cool and dry. Boil the lids for 10 minutes, then place them on a clean, lint-free towel, seal-side up, to dry. Rinse the bands and set them aside. Turn off the heat under the pot of water but keep the pot nearby until the end of this recipe, when you'll use this water bath to process the filled jars.

If you bought the yuzu at a grocery store, scrub them well in hot water. Slice off the very top of the fruit, just a thin slice to remove the stem end. Use a vegetable peeler or sharp serrated knife to remove the flavedo from the albedo, being careful to leave as much pith behind as possible, and trimming any excess pith from the back of the zest. Reserve the albedo in a small bowl. Any pith that still clings to the zest will add some bitterness to the preserve; some people will like that and some won't. Either way, the bitterness will tone down after a few weeks of storage.

Stack the zest on a cutting board. With a sharp knife, slice the zest into thin shards, and place them all in a preserving pan. Slice the yuzu in half from top to bottom, peel the core and albedo away, then chop all the albedo roughly into large chunks. Add the albedo to the reserve bowl.

Slice the fruit in half again, this time along its equator, and dig out the seeds, reserving them with the chopped albedo. Chop the fruit roughly and add it to the preserving pan with the zest. Add the water to the pan, too. (If you want to take a break, you can do so now. Cover the zest and pith, and return tomorrow.)

Tie the reserved seeds and albedo into a cheesecloth bundle and soak it with water, letting the extra water run out and giving it a quick squeeze so the bundle doesn't absorb too much water when added to the marmalade pot. The albedo and seeds from 3 pounds of yuzu makes a pretty big bag; do not be alarmed. Note that some citrus have more seeds than others. The marmalade will set with or without a lot of seeds. With a lot of seeds, it may even set quite hard. If you have very seedy citrus, use half the seeds.

Bring the zest, pulp, and water to a gentle boil, then nestle the bag of pith into it. Continue to boil with the lid on but slightly askew for 30 minutes, stirring occasionally to make sure the peels at the top are cycled to the bottom. Check to make sure the peels are cooked all the way through. If they are slightly opaque in the middle, continue boiling them until they are not.

Remove the bag, squeezing all the liquid out (hold the top of the bag with tongs and milk it with another set of tongs). Add the sugar, turn the heat to high, bring the preserves to a boil, and stir to dissolve the sugar. Do not stir further except to prevent the marmalade from boiling over. After about 18 minutes of boiling—once the preserve is bubbling madly and the surface starts to shudder instead of splash, and its color slightly darkens—check the set. The marmalade is done when, after it is dolloped onto a frozen plate and left in the freezer for 1 minute, it wrinkles when you push your finger through it. Remember to take the marmalade off the heat for the minute it takes to do this test.

Ladle the marmalade into the prepared jars, leaving ½ inch of headspace. Wipe the rims clean, place the lids, and screw the bands on, fingertip-tight (tight, but not so tight you need strong hands to reopen them). Bring the water bath back to a boil and process the jars, keeping them immersed for 10 minutes. Remove the jars from the water, and set them on the counter to cool. Store them in a dark cupboard. At first, this marmalade may be quite bitter, but that bitterness will tone down—and in many cases disappear—over time. Best eaten within 6 months. After that, citrus peels in this type of marmalade tend to toughen.

Yuzu Bath and Body Oil

The easiest way to harvest yuzu fragrance is to toss the fruit right into the bathtub for a yuzu yu, a traditional Japanese bath taken at the winter solstice. Heat and steam make the essential oils—in this case, terpenes like limonene—more volatile; volatility increases our ability to smell limonene. In a yuzu yu, fruit bobs around the body like rubber ducks, not quite the right color (more green than warm yellow) but with that same playful appeal. They are buoys of scent that raise the spirits and senses on the darkest day of the year.

This recipe captures yuzu's essential oils, but does not make what's commonly thought of as essential oil—that requires a distilling setup beyond most home cooks. Use yuzu oil as massage oil, as the oil in a sugar scrub (page 258), substitute it for vanilla oil in whipped vanilla body cream (page 299), add it to other home-made cosmetic and body products, or use as an aromatic and moisturizing addition to a hot bath.

Yield: ¾–1¼ cups

COLD EXTRACTION

This method takes 3 days and yields a fresh, bright yuzu scent.

> ½ kilogram yuzu (5 to 7 yuzu, depending on size)
> About 1½ cups sweet almond oil

With a vegetable peeler, peel the zest in wide strips from the yuzu, leaving behind as much pith (albedo) as possible. Place the zest in a clear glass container, and cover with the sweet al-mond oil. Cover the jar with a lid, and let it sit in a sunny corner. Every now and then, shake the container. If the peels aren't sub-merged in the oil after a good shake, make sure to press them back down beneath the surface. After 3 days, strain the peels from the oil. Wring the peels for the last drops of oil. Bottle the oil, and store it in a dark cupboard until you're ready to use it. Keeps indefinitely.

I found that infusing the yuzu peels in oil for longer than 3 days yielded a sweeter scent. I preferred it sharper, more sour, but both perfumes are terrific.

Make the most of the denuded yuzu by reserving their juice for sauces and saving the albedo skeletons to help thicken medlar jelly (page 178) or any other low-pectin jelly. For me, medlar and yuzu seasons intersect right around the end of November, but this might not be true elsewhere. If your timing isn't quite right for medlar jelly, freeze the yuzu skeletons to use later with something else.

HOT EXTRACTION

Yield: about ½ cup

Heat alters yuzu perfume a bit, dimming the pine scent and adding a cooked lemon scent, but it produces a scented oil quickly.

> ½ kilogram yuzu (5 to 7 yuzu, depending on size)
> ¾ cup sweet almond oil, or more as needed

With a vegetable peeler, peel the zest in wide strips from the yuzu, leaving behind as much albedo as possible. Combine the yuzu zest and sweet almond oil in a small nonreactive metal bowl. If the zest isn't completely covered by the oil, add more. Prepare a water bath by heating water in a deep saucepan and placing the bowl of almond oil and zest in the water. The water should be steaming but not boiling. Heat for 3 hours, then strain the peels from the oil and discard them. Bottle the oil, and keep it at room temperature.

Z: Zucchini

Cucurbita pepo
Cucurbitaceae (gourd) family
Also known as courgette, summer squash, baby marrow

Zucchini flowers open in the morning and close by noon. You can pick them at any time, but trying to open a closed zucchini flower is a delicate, difficult task, frustrating to a degree that it will seem expedient to take the short route and rip it open along one side so you can stuff it with goat cheese and herbs. I can tell you, having lost my patience this way, that the loss is merely aesthetic.

I planted two zucchini last year, one too many. Any zucchini is too many zucchini, Sam says. The recipe for this chapter should be "Stop Pretending There Are Recipes That Rescue Zucchini."

I like zucchini. I like knowing that of all the seeds we've planted, this one will definitely overwhelm us with our success. I plant two on our last day of average frost in the raised beds my father built over a cinder-block Jacuzzi foundation in our backyard. Each morning after, while Sam makes our coffee, I slip outside in my nightgown and bare feet to inspect the soil, see what's up.

Our backyard is small, with shared wooden fences on either side that divide us from our neighbors' equally small yards. To the west is Earl's island of green. To the north, a rutted alley. To the east, a mud pit where the neighbor's German shepherd runs in circles all day. Max hears me on the garden path and lunges at the fence, barking his head off, until he remembers the sound of my voice. "Max, it's me!" I say, as I say every day. "*Max*, for chrissakes!" Once he calms down, I hold my hand out for him to lick and nuzzle, hoping that tomorrow will be the day this gesture helps the dog cease to guard my own garden from me.

As each new sprout emerges from dirt, so does a new threat. Cutworms, onion maggots, aphids. Weeds that resemble food plants and trick me into nurturing them. Too many sprouts too close together, demanding a choice between which will stay in the ground and which will become compost. Powdery mildew. Rust. And my archnemeses, the earwigs. They hide in the folds of rhubarb leaves, between daisy petals, beneath hollyhock buds. They emerge at night to chew everything tender to the ground. I wage war on them with diatomaceous earth until I read that this white powder might also kill bees. Then I pour soy sauce and cheap cooking oil into canning jars and bury the jars in garden dirt up to the rim; earwigs, attracted to the savory scent, wander in and drown. When they multiply faster than my traps can handle, I kill them by hand.

I need Sam for this part of the battle. After dark, we take a flashlight and a wad of paper towels to the garden and hunt for earwigs—thin ovals the length of a paperclip, brownish-blackish-reddish, with pincers that turn salad into lace. Smashing them requires a quick snatch calibrated to spare the delicate plant beneath the bug, then a hard pinch through the paper towel. Earwigs' exoskeletons are strong. If we muffle our fingers in too much paper, they don't die.

"A garden is made of hope," wrote W. S. Merwin near the end of his life. "It cannot be proven, nor clutched, nor hurried."

Whose hope? Whose hurry? As we coax this garden into profusion and plenty, our produce aisle and our sanctuary, opportunistic organisms that lived here before us reassert themselves. I keep the earwigs in check by paying obsessive attention to their hiding places and midnight buffets, but as I pluck them from their suppers, I also crush nasturtiums, baby kales, clematis vines. Cultivating this landscape does not make me master over it. Which is what Merwin meant, I think, when he named the difference between what the gardener hopes and what she reaps.

This difference is not failure.

Fruit is an ovary that has matured into a package for seeds, with anatomy meant for another creature's nourishment and anatomy meant for its own. The seed takes its nourishment from the germ, while the pericarp (what we usually call "the fruit" of the fruit) entices animals to eat it, helping the seed within the fruit move to another location where a new plant can bloom. Like any organism, the seed wants to live. Usually, it will sprout regardless of the gardener, her skill or attention, or lack thereof. Whether it grows to maturity and health depends on the conditions of where it happens to root.

My mother has a story about a peach tree that, for a long time, was just a story. There was no specific occasion for the telling that I can remember. Probably she was trying to comfort me after I'd destroyed something with my own greed. Probably she did not mean it as a metaphor, because she does not speak in metaphor. We might have been talking about her life with my father in a town near the Hanford Nuclear Reservation before their children were born, where they lived in a tiny house near the Columbia River and one spring day watched volcanic ash rain from a black sky. They lived in this place for only two years, which means my mother's story occurs either the June before the May 1980 Mount Saint Helens eruption or the June immediately after.

In their backyard was a peach tree, planted by the previous owners and mature enough to bear a jackpot of fruit. She'd been told they needed to prune these peaches when they were small to encourage larger fruit and protect the health of the tree, but she couldn't do it—she couldn't choose one peach over another peach. Nor could she bear to waste fruit. That June, a surprise rainstorm swept through the Tri-Cities, more violent than anyone would expect for early summer. In the morning, my mother checked on the tree. It had split at the crown into two sagging halves. Hard green peaches littered the lawn. She was devastated.

What had she done?

When fruit falls, the impact bruises protective membranes, inviting the rot and decay necessary to break the flesh package down and release seeds. These peaches were too young for that; their tree was terminal. My father took the ax they used for firewood and chopped the peach tree to the ground.

It's early summer, and I'm in a mood, I can't sleep. I'm lying on a blanket I've thrown onto our back lawn, watching the sun come up, while Sam dreams upstairs. From this perspective, the fences on either side of me soar like loft walls that protect and enclose before giving way to roofs and open sky.

No—that description's too romantic. The sky is not open. It has been sliced into long triangles by the black lines that electrify our homes. A tree that ruins the lawn with suckers does look beautiful from this angle, all reach and no shade, like I'm underwater gazing up at a drift of kelp. A neighbor's ginger cat is on another neighbor's roof, watching me. Ignoring me. No animal alarm from Max, not yet. Where dandelions once rooted, mallow now grows in low-lying clumps, weedy relatives of hollyhocks whose small purple and white flowers mature into crunchy green buttons I could snack on, but don't. No fruit in the garden this week, just lettuce and kale, pea vines and zucchini flowers.

On the fence next to the zucchini, paper wasps crouch over a nest they're building a few inches from where I knocked down last summer's nest. Right now it's the size of a quarter. I'll leave them alone until they abandon it at the end of the season, when I'll knock it down again. Unlike yellow jackets and hornets, paper wasps are not aggressive. They're sort of goofy, or would be if their stingers were less menacing. When they fly, their long legs wave a beat behind their movements, a little left as they zip right. They eat pests and pollinate plants—zucchini, maybe? Whatever they're up to, the wasps and my household have come to an agreement. As much as possible, we leave each other alone.

The sky is pale now, headed for blue. I hear the long, high wail of trains as they rattle through the Spokane Funnel to Sandpoint or Seattle or Pasco or Kettle Falls, each car packed with wheat and rye and coal and oil and I don't know what else. Behind me, the zucchini flowers make sounds I can't hear as they open one by one, as bright as joy, the first three of 3,724,903.

This early in the season, the female and male flowers bloom out of synch, but I don't understand that yet. What I understand in this moment is that these bright-yellow flowers open before I am usually awake, close by the time I finish my second cup of coffee, then wither, unfertilized. The male flowers have slender stems, whereas the female flowers trumpet from the end of a tiny pre-fruit that looks exactly like what it is, a baby zucchini. After their triumphant morning, the male flowers curl and wither. The female flowers also wither. Their pre-fruits hang on a little longer, yellowing and shrinking into themselves over the next three days, until they drop from the vine to the dirt. In some circles this is called—really—a fruit abortion. I must be vigilant about gathering the aborted fruits so they don't invite earwigs.

A few more weeks will pass before these flowers settle into a fertile rhythm. We'll harvest the first zucchini in July. Then, through September, we'll harvest more than we can handle.

Zucchini are a New World plant, native to a sweep of the Americas from the South Central United States to central Argentina. *C. pepo,* the species to which zucchini belongs, has three categories: pumpkins, gourds, and squash. Pumpkins have a tough outer rind and fine-grained edible flesh. Their name comes from the Old English word "pompion," which comes from the Greek for "large melon." Gourds, from the Anglo-French "gourde" (in English, all common names for *C. pepo* were adopted after the European conquest of the Americas), are tough, bitter, and generally inedible, and have been used for millennia to make ornaments and practical vessels. Today, they're perhaps best known for decorative gourd season. "Squash" is the only New World etymology in this family, from the Narragansett word for vegetables eaten green, "asquutasquash." This category includes summer squashes—thin-skinned squash eaten immature—and winter squashes, which are left on the vine to develop the tough rind that makes them champions of winter storage. Zucchini are a summer squash that belong to the horticultural group marrow, selected from pumpkins and cultivated in Europe to produce long, clublike fruits—as opposed to crookneck squashes, which narrow and curve at the stem end into graceful, edible necks. Pick zucchini one to seven days after fertilization.

Or leave them on the vine much longer, until the rind hardens and the seeds develop tough coats that allow them to be roasted and eaten like pepitas. This will suck the plant's energy toward a few ripe fruits instead of encouraging it to produce a multitude of immature ones; the fruit of the mature zucchini will be too spongy and the rind too bitter to eat with pleasure.

Growing zucchini for seeds is a solution fully supported by the zucchini hater in my house: no more zucchini fritters, zucchini bread, zucchini pasta, zucchini chips, zucchini stir-fried, sautéed, baked, and stuffed. No more zucchini kebabs. No more ratatouille! I sneak zucchini into our meals only to watch Sam construct neat, limp piles of it at the edge of his plate.

I will admit that something unfair happens to zucchini squash on its journey to the grocery store. Some kind of washcloth-and-chores taste that homegrown zucchini doesn't have. Could it be that my zukes taste good to me because they are mine, so I forgive them their faults? No, I think freshness matters with this fruit more than most. Straight from the garden, they have a mildly pleasant flavor, vegetal and a little sweet. They're talented flavor-carriers and can bulk up a tomato sauce or a stir-fry with little cost or effort. A mandoline can slice zucchini so thin I see light through it, no stubborn sticking or squashing on the blade, no protest. They can be as short as cigars or as long as a Maglite, overwhelming in numbers but subdued on the tongue, not neutral, but not distinguished, either. Just *there*, ready to perform whatever modesties you ask of it.

"Nothing I do will last," writes Sarah Lindsay in "Zuc-chini Shofar," her wedding poem to gardeners. I remember this line in the weeks after my enthusiasm for zucchini has been blunted by their relentless fertility, when I've begun to forget them in the refrigerator until, while looking for something else, I stick my hand into a bag of zucchini slime. "What a congress of stinks!" writes Theodore Roethke of the promise of rotting plants. "It might make you want to stay alive even," writes Ross Gay of the wonder of compost. The zukes Sam doesn't eat, the zukes I don't cook—plus everything else that in decay might add life to the garden—we sling into a black box of recycled plastic that squats near Max's fence. This part of the garden requires attention, too. Not too little, not too much. Every few weeks, as the top of the compost pile dissolves into slime, I cover my nose and mouth with one of the masks we bought for fire season, glove up, and mix the pile by hand, adding fallen oak leaves I harvested from the feet of our trees last fall. Worms writhe and pill bugs scatter as I disturb their feast. It is satisfying, this mix and chew and smear and rot. We're making earth.

The yard where I grow zucchini is just a yard. Not a place I am from, but a place I am rooted. Dirt that belonged to someone else before it belonged to me, or dirt that does not belong to me, but dirt I currently bear title to, dirt I nurture, fuss, and pluck. If Sam and I have a child, that child will run here, in this grass, between fences we'll reinforce to keep the child within, to keep threats without; beneath the elder bush we planted for privacy and protection and elderberry harvests; around the quince tree we'll plant for fragrance and beauty, and maybe, when we think of it, as a reminder to contemplate the soul's marriage with God, or the natural world, or the constructed world, if between these there's a meaningful difference. We'll try to grow a garden so comfortable and sweet, so without difficulty, that the child won't notice she's been locked in. We won't be able to help ourselves—the only selves we can truly help. I hope, at the right time, we fail.

Zucchini Blossoms with Chèvre and Herbs

Start by building a garden. Clear the soil of weeds, and amend it with compost. On the last day of average frost, plant two zucchini seeds (one for your garden, one for backup), 8 inches apart. Plan for the zucchini eventually to take over about 9 square feet of the garden. Once the two zucchini are a few inches tall, dig up the smaller one and give it to a neighbor. Continue with one zucchini plant, watering and weeding as needed. It may be a month before you see your first bloom.

Zucchini blossoms can be eaten at any time during the growing season. Female zucchini blossoms are especially pretty—they can be picked, stuffed, and seared according to the directions below, and include a miniature zuke at their stem end. If you'd like to harvest the flowers throughout the season but still want zucchini fruit, pick the male flowers, run a Q-tip over their stamen and anthers, then hand-pollinate the female flowers by brushing the pollen-covered Q-tip on their stigmas.

If you pick zucchini flowers in the morning, they'll stay open on the countertop until dinner. Pinch off the stamens right before cooking, and trim the stem so it's just long enough to use as a handle as you eat the whole fruit. In a small bowl, combine cold chèvre with chopped fresh herbs (marjoram or sage or tarragon or dill—any herb will work) and a little salt. Taste, and adjust herbs and salt as needed. Then roll the herbed chèvre into a small ball (adjust the size to fit your flower), and stuff the flowers with the chèvre. Twist the petals shut (don't worry if they flop open), heat a neutral frying oil over medium-high heat until it pops, then sear the flowers briefly on both sides until they just turn toasty brown. Eat all of the stuffed flowers immediately, when they're almost too hot to handle.

Garlicky Fermented Zucchini

When faced with more zucchini than you can (or want to) eat, extend its shelf life and heighten its savory qualities by fermenting it with garlic and herbs. Serve with cheese, cured meats, on sandwiches, as salsa, as a dip—wherever a spicy, garlicky, herby condiment would be welcome. Reserve the brine created by this recipe to start more vegetable ferments (replace the zukes I call for here with eggplant, for example), use it as vegetable stock for a cold summer soup, or add it to Bloody Marys.

Yield: about 8 cups

1 kilogram (2.2 pounds) zucchini, cut into ¼-inch cubes (do not peel)

1 or 2 sweet red peppers, seeded, cut into ¼-inch dice (for color)

Hot peppers, any kind, minced (as many or as few as you like)

1 small head of garlic, peeled and crushed or minced

Salt, to taste (enough so the zukes taste good but are not overwhelmingly salty)

OPTIONAL SPICES AND HERBS

About 1 teaspoon coriander ground in a mortar

About 1 teaspoon cumin ground in a mortar

1 handful chopped fresh herbs (oregano, thyme, lovage, whatever sounds good)

Combine all the ingredients in a large bowl. Taste, and add more salt if you'd like it saltier. Squeeze the mixture by hand to help coax out the brine. It may not release much yet. Pack the mixture into a large, lidded container (one 2-quart or two 1-quart jars will do nicely), pressing the vegetables down as you fill the jar to eliminate air pockets. Leave plenty of headspace. Weigh the vegetables down (I like to use the plastic top from a yogurt pint, cut once from edge to middle so it can be bent into a cone shape and used to wedge the vegetables down), and secure the jar's lid. If there isn't much brine yet, return in 1 day, once the salt has drawn more water from the vegetables, and tamp the veggies below the brine. Leave the jars in a warm spot for 3 to 7 days, removing their tops once a day to relieve the gases that build up during fermentation. Taste the zucchini every now and then. When it's sour enough (that's up to you), drain off the brine, jar the fermented zukes, tamp them down to reduce air pockets, then cap with high-quality olive oil and store in the refrigerator. The olive oil will extend the vegetables' shelf life. And it tastes good.

I like this best after about 4 days of fermentation, when the zucchini is still a little crisp.

Sources and Recommended Reading

Alexander, Susan J., and Rebecca T. Richards. *A Social History of Wild Huckleberry Picking in the Pacific Northwest*. Portland, Ore.: U.S. Department of Agriculture, Forest Service, Pacific Northwest Research Station, 2006.

Annwen. *Herbal Abortion: A Woman's D.I.Y. Guide*. Godhaven Ink, 2002.

Arber, Agnes Robertson. *Herbals, Their Origin and Evolution: A Chapter in the History of Botany, 1470–1670*. Cambridge Science Classics. 3rd ed. Cambridge and New York: Cambridge University Press, 1986.

Baker, Kate. *Captured Landscape: The Paradox of the Enclosed Garden*. London and New York: Routledge, 2012.

Bangsberg, P. T. "China Airlines Finds New Logo Won't Fly." *The Journal of Commerce and Commercial Bulletin* 406, no. 28583 (1995).

Beyerl, Paul. *The Master Book of Herbalism*. Custer, Wash.: Phoenix Publishing, 1984.

Bowen, 'Asta. *The Huckleberry Book*. Helena, Mont.: American Geographic, 1988.

Boyd, Robert, ed. *Indians, Fire and the Land in the Pacific Northwest*. Corvallis: Oregon State University Press, 1999.

Boyd, Robert. *People of the Dalles: The Indians of Wascopam Mission*. Lincoln: University of Nebraska Press, 1996.

Briggs, Raleigh. *Make Your Place: Affordable, Sustainable Nesting Skills*. Portland, Ore.: Microcosm Publishing, 2008.

Burbank, Luther. *The Training of the Human Plant*. New York: Century Co., 1907.

Caffrey, Cait. "Superfoods." *Salem Press Encyclopedia of Health*, 2019. Research Starters, EBSCO.

Chekhov, Anton. *The Portable Chekhov*. Edited by Avrahm Yarmolinsky. Viking Portable Library. New York: Penguin, 1978.

Child, Francis James, ed. *English and Scottish Popular Ballads*. Vol. 3. 1860. Reprint, New York: Dover Publications, Inc., 1965.

Collingham, Lizzie. *The Taste of Empire: How Britain's Quest for Food Shaped the Modern World*. New York: Basic Books, 2017.

Conrad, CA. "Mount Monadnock Transmissions." *(Soma)tic Poetry Rituals*, September 29, 2016, http://somaticpoetryexercises.blogspot.com/2016/09.

Culpeper, Nicholas. *Culpeper's Complete Herbal*. London: Arcturus, 2009.

Culpeper, Nicholas. *Complete Herbal*. Buck, England: W. Foulsham & Co., Ltd., 1975.

Dafni, Amots, and E. Lev. "The Doctrine of Signatures in Present-Day Israel." *Economic Botany* 56, no. 4 (2002), 328–34. https://doi.org/10.1663/0013-0001(2002)056[0328:TDOSIP]2.0.CO;2.

Danbom, David B. "Flour Power: The Significance of Flour Milling at the Falls." *Minnesota History* 58, no. 5/6 (2003): 271–85.

Daudet, Alphonse. *In the Land of Pain*. Vintage Classics. New York: Vintage Books, 2016.

Davidson, Alan, and Tom Jaine. *The Oxford Companion to Food*. Oxford, U.K.: Oxford University Press, 2006.

Davis, William. *Wheat Belly: Lose the Wheat, Lose the Weight, and Find Your Path Back to Health*. Emmaus, Penn.: Rodale, 2011.

Derig, Betty B., and Margaret C. Fuller. *Wild Berries of the West*. Missoula, Mont.: Mountain Press, 2001.

Deur, Douglas. "A Most Sacred Place: The Significance of Crater

Lake Among the Indians of Southern Oregon." *Oregon Historical Quarterly* 103, no. 1 (2002): 18–49.

Dutton, Julian. *Keeping Quiet: Visual Comedy in the Age of Sound.* Gosport, U.K.: Chaplin Books, 2015.

Edwards, Nina. *Weeds.* London: Reaktion Books, 2015.

European Food Safety Authority Panel on Contaminants in the Food Chain. "Acute health risks related to the presence of cyanogenic glycosides in raw apricot kernels and products derived from raw apricot kernels," *EFSA Journal* 14, no. 4 (2016). https://doi.org/10.2903/j.efsa.2016.4424.

Ferber, Christine. *Mes Confitures: The Jams and Jellies of Christine Ferber.* East Lansing: Michigan State University Press, 2002.

Field Guide for Managing Himalayan Blackberry in the Southwest. Washington, D.C.: USDA Forest Service, Feb. 2015. https://www.fs.usda.gov/Internet/FSE_DOCUMENTS /stelprd3828954.pdf.

Fortini, Amanda. "Pomegranate Princess." *The New Yorker* 84, no. 7 (2008).

Foust, Clifford M. *Rhubarb: The Wondrous Drug.* Princeton, N.J.: Princeton University Press, 1992.

Frazer, James George, and Robert Fraser. *The Golden Bough: A Study in Magic and Religion.* World's Classics. Oxford, U.K., and New York: Oxford University Press, 1994.

Frey, Emily Kendal. *Sorrow Arrow.* Portland, Ore.: Octopus Books, 2014.

Gabriel, Julie. *The Green Beauty Guide: Your Essential Resource to Organic and Natural Skin Care, Hair Care, Makeup, and Fragrances.* Deerfield Beach, Fla.: Health Communications, 2008.

Gafner, Stefan. "Scientific Journals Increasingly Skeptical of Antioxidant Research." *HerbalGram*, no. 117 (2018): 35.

Gay, Ross. *Catalog of Unabashed Gratitude.* Pitt Poetry Series. Pittsburgh, Pa.: University of Pittsburgh Press, 2015.

Gerard, John, and Marcus Woodward. *Gerard's Herbal: The History of Plants.* London: Senate, 1994.

Gibbons, Euell. *Stalking the Wild Asparagus.* Twenty-fifth anniversary edition. Putney, Vt.: Alan C. Hood, 1987.

Gilette, F. L., and Hugo Zieman. *The White House Cook Book.* New York, Akron, and Chicago: The Saalfield Publishing Company, 1905.

Goldstein, Darra, ed. *The Oxford Companion to Sugar and Sweets.* Oxford, U.K., and New York: Oxford University Press, 2015.

Gould, H. P. *The Himalaya Blackberry.* Bureau of Plant Industry. Circular 116. Issued March 8, 1913.

Grieve, M. *A Modern Herbal: The Medicinal, Culinary, Cosmetic and Economic Properties, Cultivation and Folklore of Herbs, Grasses, Fungi, Shrubs and Trees with All Their Modern Scientific Uses.* Edited by Mrs. C. F. Leyel. London: Cape, 1974.

Grigson, Jane. *Jane Grigson's Fruit Book.* Harmondsworth, U.K.: Penguin, 1982.

Gunther, Erna. *Ethnobotany of Western Washington.* 1945. Rev. ed., Seattle: University of Washington Press, 1973.

Hachisu, Nancy Singleton. *Preserving the Japanese Way: Traditions of Salting, Fermenting, and Pickling for the Modern Kitchen.* Kansas City, Mo.: Andrews McMeel Publishing, 2015.

Hafezi, F., et al. "*Actinidia Deliciosa* (Kiwifruit), a New Drug for Enzymatic Debridement of Acute Burn Wounds." *Burns* 36, no. 3 (2010): 352–55. https://doi.org/10.1016/j.burns.2009.04.021.

Hamilton, Edith. *Mythology: Timeless Tales of Gods and Heroes.* New York: Penguin, 1940.

Hammers, Mia, et al. "Constructing a Genetic Linkage Map of *Vitis Aestivalis*-Derived 'Norton' and Its Use in Comparing Norton and Cynthiana." *Molecular Breeding* 37, no. 5 (2017): 1–14. https://doi.org/10.1007/s11032-017-0644-6.

Hart, Linda, and LeAnna DeAngelo, Ph.D. "Antioxidants." *Magill's Medical Guide (Online Edition),* 2018. Research Starters, EBSCO.

Hatfield, Gabrielle. *Encyclopedia of Folk Medicine: Old World and New World Traditions.* Santa Barbara, Calif.: ABC-CLIO, 2004.

Hedrick, U. P. *Cyclopedia of Hardy Fruits.* New York: Macmillan, 1922.

Hempel, Amy. *Reasons to Live: Stories.* New York: Penguin, 1986.

"Himalayan Blackberry." *USDA Plant Guide,* January 2014.

Hohman, John George. *Pow-Wows; or, The Long Lost Friend: famous Witchbook of the Pennsylvania Dutch*. 1820. Reprint, American History Publications, 1971.

Holy Bible. Edited by Stephen J. Hartdegen and Christian P. Ceroke. Charlotte, N.C.: C.D. Stampley Enterprises, Inc., 1970.

Hummer, Kim E., et al. "Luther Burbank's Best Berries." *Hortscience* 50, no. 2 (2015): 205–10.

Itoh, Makiko. *The Just Bento Cookbook: Everyday Lunches to Go*. Tokyo: Kodansha International, 2010.

James, Glyn, and Frank Blackburn. *Sugarcane*. Oxford, U.K.: Blackwell Science, 2004.

Jeunet, Catherine Marie. *Reclaiming Our Ancient Wisdom: Herbal Abortion Procedure and Practice for Midwives and Herbalists*. Portland, Ore.: Eberhardt Press, 2016.

Kallas, John. *Edible Wild Plants: Wild Foods from Dirt to Plate*. Wild Food Adventure Series. Layton, Utah: Gibbs Smith, 2010.

Katz, Sandor Ellix. *Wild Fermentation: The Flavor, Nutrition, and Craft of Live-Culture Foods*. 2nd ed. White River Junction, Vt.: Chelsea Green Publishing, 2016.

Kimmerer, Robin Wall. *Braiding Sweetgrass*. Minneapolis: Milkweed Editions, 2013.

Kindscher, Kelly. *Edible Wild Plants of the Prairie: An Ethnobotanical Guide*. Lawrence: University Press of Kansas, 1987.

Kliman, Todd. *The Wild Vine: A Forgotten Grape and the Untold Story of American Wine*. New York: Clarkson Potter, 2010.

Kwasny, Melissa. *Thistle: Poems*. Sandpoint, Idaho: Lost Horse Press, 2006.

Langley, P. "Why a Pomegranate?" *British Medical Journal* 321 (2000): 1153–54. https://doi.org/10.1136/bmj.321.7269.1153.

Lanza, Anna Tasca. *The Garden of Endangered Fruit*. Palermo, Italy: Stamperia Zito, 2004.

Laszlo, Pierre. *Citrus: A History*. Chicago: University of Chicago Press, 2007.

Li, Yanmei, and Qinglin Liu. "*Prunus Mume*: History and Culture in China." *Chronica Horticulturae* 51, no. 3 (2011): 28–35. https://www.actahort.org/chronica/pdf/ch5103.pdf#page=28.

Lin, Lin, et al. "Xylitol Nasal Irrigation in the Treatment of Chronic Rhinosinusitis." *American Journal of Otolaryngology* 38, no. 4 (2017): 383–89. https://doi.org/10.1016/j.amjoto .2017.03.006.

Lindsay, Sarah. *Twigs & Knucklebones*. Port Townsend, Wash.: Copper Canyon Press, 2008.

Louis, M. K. "Proserpine and Pessimism: Goddesses of Death, Life, and Language from Swinburne to Wharton." *Modern Philology* 96, no. 3 (1999): 312–46. https://doi.org/10.1086 /492763.

Lubinsky, Pesach, et al. "Origins and Dispersal of Cultivated Vanilla (*Vanilla planifolia* Jacks. [Orchidaceae])." *Economic Botany* 62, no. 2 (2008): 127–38.

Lust, John B. *The Herb Book*. New York: Benedict Lust Publications, 1974.

Mahood, M. M. *The Poet as Botanist*. Cambridge, U.K.: Cambridge University Press, 2008.

Maloney, Mika. *A Small Batch*. Spokane, Wash.: self-published, 2019.

Marcel, C. "Pomegranate." *CINAHL Nursing Guide EBSCO Publishing*, April 2018. Nursing Reference Center Plus, EBSCO.

Matsumoto, Kōsai, II. *Traditional Herbs for Natural Healing*. Osaka: Kosai Matsumoto Office, 1977.

McPhee, John. *Oranges*. New York: Farrar, Straus and Giroux, 1967.

Meech, William Witler. *Quince Culture: An Illustrated Handbook for the Propagation and Cultivation of the Quince, with Descriptions of Its Varieties, Insect Enemies, Diseases and Their Remedies*. New York: Orange Judd, 1888.

Mekonnen, Serkalem. "I Swallowed a Cherry Pit! Are Stone Fruit Pits Poisonous?" *Poison Control*, May 2, 2020. https://www .poison.org/articles/i-swallowed-a-cherry-pit-184.

Merwin, W. S., and Larry Cameron. *What Is a Garden?* Columbia: University of South Carolina Press, 2016.

Mintz, Sidney W. *Sweetness and Power: The Place of Sugar in Modern History*. New York: Penguin, 1986.

Miranda, Carolina A. "Q&A: Kara Walker on the Bit of Sugar

Sphinx She Saved, Video She's Making." *Los Angeles Times,* October 13, 2014.

Mirel, Elizabeth Post. *Plum Crazy: A Book About Beach Plums.* New York: C. N. Potter, 1973.

Moerman, Daniel E. *Native American Ethnobotany.* Portland, Ore.: Timber Press, 1998.

———. *Native American Medicinal Plants: An Ethnobotanical Dictionary.* Portland, Ore.: Timber Press, 2009.

Mudge, Zachariah Atwell. *Sketches of Mission Life Among the Indians of Oregon.* New York: Carlton & Porter, 1854.

Munari, Bruno. *Good Design.* 1963. Reprint, Mantova, Italy: Corraini Edizioni, 1997.

Opfer, Chris. "The Norton Grape: American Viticulture's Native Son." *Gastronomica* 11, no. 3, University of California Press (2011): 92–95.

Organizing Committee, 24th International Horticultural Congress, ed. *Horticulture in Japan.* Tokyo: Asakura Publishing, 1994.

Patel, Shailja. "Unpour." *Creative Time Reports,* May 8, 2014. https://creativetimereports.org/2014/05/08/shailja-patel-unpour-kara-walker-domino-sugar-factory.

Rain, Patricia. *Vanilla: The Cultural History of the World's Favorite Flavor and Fragrance.* New York: Jeremy P. Tarcher/Penguin, 2004.

Redzepi, René, and David Zilber. *The Noma Guide to Fermentation: Foundations of Flavor.* New York: Artisan, 2018.

Reich, Lee. *Uncommon Fruits for Every Garden.* Portland, Ore.: Timber Press, 2004.

"Rhubarb (Alternative Therapy)." *Health Library: Evidence-Based Information EBSCO Publishing,* December 2015. Nursing Reference Center Plus, EBSCO.

"Rhubarb as a Dietary Supplement." *Salem Press Encyclopedia of Health,* 2017. Research Starters, EBSCO.

Riddle, John M. *Eve's Herbs: A History of Contraception and Abortion in the West.* Cambridge, Mass.: Harvard University Press, 1997.

Ritsos, Yannis. *The Fourth Dimension*. Princeton, N.J.: Princeton University Press, 1993.

Roethke, Theodore. *Words for the Wind*. Garden City, N.Y.: Doubleday, 1958.

Ruthnum, Naben. *Curry: Eating, Reading, and Race*. Toronto: Coach House Books, 2017.

Simpler, Simon the. *An Herbal Medicine-Making Primer*. Zine.

Slater, Nigel. *Ripe: A Cook in the Orchard*. 1st Ten Speed Press edition. Berkeley, Calif.: Ten Speed Press, 2012.

Small, Ernest. *North American Cornucopia: Top 100 Indigenous Food Plants*. 1st ed. Boca Raton, Fla.: CRC Press, Taylor & Francis Group, 2013.

——. *Top 100 Exotic Food Plants*. Boca Raton, Fla.: CRC Press, 2012.

Smith, Jane S. *The Garden of Invention: Luther Burbank and the Business of Breeding Plants*. New York: Penguin Press, 2009.

Solmonson, Lesley Jacobs. *Gin: A Global History*. London: Reaktion Books, 2012.

Sonneman, Toby F. *Lemon: A Global History*. London: Reaktion Books, 2012.

Sontag, Susan. *Illness as Metaphor, and AIDS and Its Metaphors*. New York: Doubleday, 1990.

Sõukand, Renata, and Raivo Kalle. "Herbal Landscape: The Perception of Landscape as a Source of Medicinal Plants." *Trames: A Journal of the Humanities and Social Sciences* 14, no. 3 (2010): 207–26. https://doi.org/10.3176/tr.2010.3.01.

Specter, Michael. "Miracle in a Bottle." *The New Yorker*, February 2, 2004.

Stein, Gertrude. *Tender Buttons: Objects, Food, Rooms*. New York: Claire Marie, 1914.

Stevenson, Angus, and Christine A. Lindberg. *New Oxford American Dictionary*. Oxford, U.K., and New York: Oxford University Press, 2010.

Stewart, Amy. *The Drunken Botanist: The Plants That Create the World's Great Drinks*. Chapel Hill, N.C.: Algonquin Books of Chapel Hill, 2013.

Stockwell, Christine. *Nature's Pharmacy: A History of Plants and Healing*. London: Century, 1988.

Sturtevant, Ernest A., ed. *History of the Great Mill Explosion at Minneapolis, May 2, 1878, with a View of the Ruins After the Fire*. Minneapolis, Minn.: Davison & Hart, 1878.

Sumner, Judith. *American Household Botany: A History of Useful Plants, 1620–1900*. Portland, Ore.: Timber Press, 2004.

Taillac, Victoire de, and Ramdane Touhami. *An Atlas of Natural Beauty: Botanical Ingredients for Retaining and Enhancing Beauty*. New York: Simon & Schuster, 2018.

Tanis, David. *A Platter of Figs and Other Recipes*. New York: Artisan, 2008.

Trefis Team, "Another L Brands Story: The Ingredients Behind the Sweet Smell of Bath & Body Works' Success." *Forbes*, September 15, 2015. https://www.forbes.com/sites/greatspeculations/2015/09/21/another-l-brands-story-the-ingredients-behind-the-sweet-smell-of-bath-body-works-success/#46bdb79a3851.

Vancouver, George. *A Voyage of Discovery to the North Pacific Ocean, and Round the World*. London: G. G. and J. Robinson, Paternoster-Row, and J. Edwards, Pall-Mall, 1798.

Van der Zee, Barbara. *Green Pharmacy: The History and Evolution of Western Herbal Medicine*. Rochester, Vt.: Healing Arts Press, 1997.

Vaughan, J. G., and C. A. Geissler. *The New Oxford Book of Food Plants*. Oxford, U.K.: Oxford University Press, 1997.

Wallis, Faith. *Medieval Medicine: A Reader*. Readings in Medieval Civilizations and Cultures. Toronto: University of Toronto Press, 2010.

Welsch, Roger L., and Linda K. Welsch. *Cather's Kitchens: Foodways in Literature and Life*. Lincoln: University of Nebraska Press, 1987.

West, Kevin. *Saving the Season: A Cook's Guide to Home Canning, Pickling, and Preserving*. New York: Alfred A. Knopf, 2013.

Wharton, Edith. "Pomegranate Seed." *Scribner's Magazine* 51 (1912). ProQuest.

Wilson, Edward O. *Biophilia*. Cambridge, Mass.: Harvard University Press, 1984.

Woolf, Virginia. *A Room of One's Own*. 1st Harvest/HBJ edition. San Diego: Harcourt Brace Jovanovich, 1989.

Yoon, Carol Kaesuk. *Naming Nature: The Clash Between Instinct and Science*. New York: W. W. Norton, 2009.

Young, Kay. *Wild Seasons: Gathering and Cooking Wild Plants of the Great Plains*. Lincoln: University of Nebraska Press, 1993.

Zucker, Rachel. *Eating in the Underworld*. Wesleyan Poetry. Middletown, Conn.: Wesleyan University Press, 2003.

Acknowledgments

My deep thanks to those who read drafts, provided essential conversation, sent pounds of difficult fruit, and tested my recipes: Argyle Baukol, Brian Blair, Raleigh Briggs, Kelly Chadwick, Vicki Croft, Piper J. Daniels, Jim Demetre, Katherine Eulensen, Dr. Jennifer Fang, Tony Flinn, Sam Foley, Evelyn Franz, Emily Kendal Frey, CMarie Fuhrman, Nina Mukerjee Furstenau, Terry Furstenau, Knox Gardner, Whitney Jacques, Kim Kent, Doug Lebo, Heather Malcolm, Mika Maloney, Chelsea Martin, Summer Miller, Rick Misterly, Frieda Morgenstern, Lindsey Mutschler, Kathryn Nuernberger, Suzanne Paola, Marie Prado, Laura Read, Cinda Reed, Sharma Shields, Kathryn Smith, Alexandra Teague, Kelen Tuttle, Elissa Washuta, Kary Wayson, Ellen Welcker, Lisa Wells, Katie Yamashita, Maya Jewell Zeller, and Ellen Ziegler.

Special thanks to Lora Lea Misterly for lending her wisdom and curiosity to my difficult fruit experiments; to Paul Fang for the beautiful translation of Xie Xie's "Early Plum"; and to Yuka Igarashi for publishing an early version of "C: Cherry" in *Catapult*.

I am grateful to the many people who let me interview them for this book, including those not otherwise named in the text: Shin Yu Pai, Barb Burrill and Natalie Place of City Fruit, Melissa

Ehrman of the Midwest Aronia Association, Dr. Ai Hisano of Kyoto University, Jacob Hollman and Drew Lemberger of Les Bourgeois Vineyards, and Dr. Bernadine Strik of Oregon State University. Thank you to the Mill City Museum in Minneapolis, Minnesota, for assistance while researching the history of the Washburn A Mill, and to Rebecca Alexander at the University of Washington's Botanic Gardens and Miller Library in Seattle.

Emma Paterson's guidance, guts, and grace made me believe these difficult fruits could become the book I'd imagined. Thank you.

Jenna Johnson and Ansa Khan Khattak are superb editors who stuck with me through my unruly writing process and made this book better. Thank you.

To my family for their unconditional love and support: thank you.

Thank you most of all to Sam.

Index

Page numbers in *italics* refer to illustrations.

for mother, 9–10, 12–13, 14
of other people, 315, 326
treatment, 12–14, 324–27
unknown cause of, 321
paper dye, 16–17
parsley
gremolata, 80–82
peach tree, 348–49
pectin
apple, 148–49
in gooseberries, 87
lemon, 178
preserves and, 87, 339
quince and, 221–22
sugar and, 250
yuzu, 178, 334
People of the Dalles (Boyd), 99
peppermints, xylitol, 330
peppers, 356–57
perfumes
bitter-almond extracts in, 296–97
vanilla, 287–89, 291–93, 295–99
Persephone, 207–208, 210, 212–14
persipan, 34
cherry pie and, 36
recipe, 126–27
pesticides, 119
pharmacy, 57, 58
Physical Directory, A (Culpeper), 327

physical therapy, 12, 153–59, 160
pickled rhubarb, 239
pie
cherry, 35–38
gooseberry, 86, 88–89
huckleberry, 110–11
Iowa State Fair contest for, 86, 88–89
making, 35–38, 86, 88–89, 307–8, 317
pastry recipe, 317
pits, *see* seeds/pits
placebo effect, 59–60, 260
plant breeding, 23–24
Platter of Figs and Other Recipes, A (Tanis), 80
Plum Crazy (Mirel), 28
plums, *see* Italian plums; shiro plums; ume plums
poison, *see* toxicity
pomegranates (*Punica granatum*), 207
fertility and, 210–11, 213
Greek mythology around, 207–208, 209, 210, 212–14
health benefits, 209, 210–11, 217
juice, 209–10, 217
life and death cycle and, 210, 212, 214
molasses, 215–16
religious lore about, 208–209
semi-deathless pomegranate mask, 217

sourcing, 282

ume plums compared with, 272

shiso

about, 48, 276–77, 279, 282, 285

chicken katsu rolls with umeboshi and shiso, 285–86

umeboshi/shiroboshi, 282–84

shrub (beverage), 28–29

sinus wash, 329

Slater, Nigel, 63

slavery, 190, 243, 245, 249, 256, 290

Small, Ernest, 141

Small Batch, A (Maloney), 301

smoothie, aronia, 4–5, 15

Sorrow Arrow (Frey), 211

soups

faceclock greens, fennel sausage, and barley soup, 80–82

spider balls, 205

spring fever, 211

Spokane

farmers' markets in, 103, 174

river and falls, 212

Salish language preservation and restoration in, 107–109

search for durian in, 47, 52

search for huckleberries in, 103, 106,

trains through, 352

wildfires near, 103–104, 269–71, 274, 276–277, 280

Staffel, Rebecca, 250, 251

Stalking the Wild Asparagus (Gibbons), 77, 133

Stewart, Martha, 199–200

stone fruit. *see also specific stone fruits*

"almond" extract from, 32–34, 36, 40–42, 43, 121, 126–27

pits and toxicity, 33–34, 38, 40–41, 126, 127, 280

Stubecki, Lou, 119

Subtlety, or the Marvelous Sugar Baby, A (Walker), 245–48, 252, 254–57

sugar alcohol, *see* xylitol

sugarcane (*Saccharum officinarum*), *243*

art installation and, 245–48, 252, 254–57

cane sugar scrub, 258–59

consumption statistics, 244

cutting and biting into, 251–52

factory, 245–48, 253, 256–57

fruit relationship with, 249–50

growing and harvest, 245, 248

jam ratios of fruit to, 250–51

origins and history, 243–45